Real Project Planning

Real Project Planning
Developing a Project Delivery Strategy

Dr Trish Melton

AMSTERDAM • BOSTON • HEIDELBERG • LONDON •
NEW YORK • OXFORD • PARIS • SAN DIEGO •
SAN FRANCISCO • SINGAPORE • SYDNEY • TOKYO
Butterworth-Heinemann is an imprint of Elsevier

Butterworth-Heinemann is an imprint of Elsevier
Linacre House, Jordan Hill, Oxford OX2 8DP, UK
30 Corporate Drive, Suite 400, Burlington, MA 01803, USA

1st Edition 2008

Notice
No responsibility is assumed by the publisher for any injury and/or damage to persons or property as a matter
of products liability, negligence or otherwise, or from any use or operation of any methods, products, instructions
or ideas contained in the material herein. Because of rapid advances in the medical sciences, in particular,
independent verification of diagnoses and drug dosages should be made

British Library Cataloguing in Publication Data
A catalogue record for this book is available from the British Library

Library of Congress Cataloging-in-Publication Data
A catalog record for this book is available from the Library of Congress

ISBN: 978-0-7506-8472-9

For information on all Butterworth-Heinemann publications
visit our web site at books.elsevier.com

Typeset by Charon Tec Ltd (A Macmillan Company), Chennai, India
www.charontec.com
Printed and bound in Great Britain

07 08 09 10   10 9 8 7 6 5 4 3 2 1

About the author

Trish Melton is a project and business change professional who has worked on engineering and non-engineering projects worldwide throughout her career. She works predominantly in the chemicals, pharmaceuticals and healthcare industries.

She is a chartered Chemical Engineer and a Fellow of the Institution of Chemical Engineers (IChemE), where she was the founder Chair of the IChemE Project Management Subject Group. She is a part of the Membership Committee which reviews all applications for corporate membership of the institution and in 2005 she was elected to the Council (Board of Trustees).

She is an active member of the International Society of Pharmaceutical Engineering (ISPE) where she serves on the working group in charge of updating ISPE's *Bulk Pharmaceutical Chemicals Baseline® Guide*. She is the founder and Chair of the Project Management Community of Practice formed in 2005. She has presented on various subjects at ISPE conferences including project management, quality risk management and lean manufacturing, and has also supported ISPE as the conference leader for project management and pharmaceutical engineering conferences. She is also the developer and lead trainer for ISPE's project management training course. In 2006 the UK Affiliate recognized Trish's achievements when she was awarded their Special Member Recognition Award.

Trish is the Managing Director of MIME Solutions Ltd., an engineering and management consultancy providing project management, business change management, regulatory, and GMP consulting for pharmaceutical, chemical and healthcare clients.

Within her business, Trish is focused on the effective solution of business challenges and these inevitably revolve around some form of project: whether a capital project, an organizational change programme or an interim business solution. Trish uses project management on a daily basis to support the identification of issues for clients and implementation of appropriate, sustainable solutions.

Good project management equals good business management and Trish continues to research and adapt best practice project management in a bid to develop, innovate and offer a more agile approach.

About the Project Management Essentials series

The Project Management Essentials series comprises four titles written by experts in their field and developed as practical guidelines, suitable as both university textbooks and refreshers/additional learning for practicing project managers:

- Project Management Toolkit: The Basics for Project Success.
- Project Benefits Management: Linking Projects to the Business.
- Real Project Planning: Developing a Project Delivery Strategy.
- Managing Project Delivery: Maintaining Control and Achieving Success.

The books in the series are supported by an accompanying website http://books.elsevier.com/companions, which delivers blank tool templates for the reader to download for personal use.

Foreword

This book has become a reality for a number of reasons:

- As an experienced Project Manager I realized that more and more I was dealing with customers, sponsors and Project Team members who had no project management experience. My first book in this series: *Project Management Toolkit* (Melton, 2007), was a direct response to that. However I have found that project delivery planning is a particular area where expertise is lacking.
- As the founder Chair of the IChemE Project Management Subject Group (PMSG) and then more recently a part of the Continuous Professional Development (CPD) and Publications Sub-groups it was also evident that there wasn't a full series of books which would support the further development of the Project Manager.

Real Project Planning: Developing your Project Delivery Strategy is intended to be a more in-depth look at the second value-added stage in a project and builds on from Chapter 4 of the *Project Management Toolkit* (Melton, 2007).

The other books in the project management series, (Melton, 2007) are outlined earlier (page vii).

Although this book is primarily written from the perspective of engineering projects within the process industries, experiences from both outside of this industry and within different types of projects have been used.

The tools, methodologies and examples are specific enough to support engineering managers developing project delivery plans for projects within the process industries yet generic enough to support the R&D manager in developing a plan to develop or launch a new product; the business manager in planning to transform a business area or the IT manager in planning to deliver a new computer system. The breadth of the short and full case studies demonstrate the generic use of the planning methodologies presented over a wide range of industries and project types.

Project management is about people and this book will emphasize the criticality of the development of plans to support the 'soft' side of projects: the people whose lives may change as a result of a project, the Project Team members who are key to effective delivery, the sponsors and organizational stakeholders who ensure, with the Project Manager, that 'no project is an island'.

Acknowledgements

In writing a book which attempts to share a greater level of expertise than previously (*Project Management Toolkit*), you need to develop that expertise, gain peer review and then share and test it. I therefore want to acknowledge a number of people against these specifics:

For supporting the development of my project delivery planning expertise over many years:

➤ Ray Scherzer, GSK.

For supporting the peer review of this collated project delivery planning expertise:

➤ Arnold Black (Member of the IChemE Project Management Subject Group Committee).
➤ Bill Wilson, Paul Burke and Jeff Wardle, AstraZeneca.

For sharing and testing this collated project delivery planning expertise on real 'live' projects:

➤ All my current and past clients.
➤ Associates of MIME Solutions Ltd such as Victoria Bate, Andrew Roberts and others.

Author's Note: Although all the case studies presented in this book are based on real experiences they have been suitably altered so as to maintain complete confidentiality.

How to use this book

When you pick up this book I am hoping that before you delve into the content you'll start by glancing here.

The structure for the book is based around the concept that every project goes through three types of planning phases – these are described in Chapter 1 and then each phase becomes the subject of its own chapter (2 to 4).

Chapter 1 is a general introduction to the concept of building a robust Project Delivery Plan – a term which conveys so much more than simply 'planning'. This can be read at any time to refresh you on some basic concepts which are applied within the core chapters. This chapter also provides the link between the *Project Management Toolkit* (Melton, 2007) and this more in-depth look at the second value-added stage in a project.

Chapters 2 to 4 are the 'core chapters' made up of the following generic sections:

➤ Introduction of detailed planning concepts.
➤ Presentation of specific methodologies and how they support effective planning.
➤ Introduction of planning tools and associated tool templates.
➤ Demonstration of chapter concepts, methodologies and tool use though the use of case studies.
➤ Summary of handy hints.

Each core chapter can 'stand-alone', so the reader can dip into any planning phase.

Chapter 5 pulls together all the previous planning concepts into a suggested structure for a Project Delivery Plan document and also discusses more complex programme management planning issues.

Chapters 6 to 8 contain a series of case study projects, and in effect, are the culmination of the use of all planning concepts and methodologies introduced in the previous chapters. These aim to show the breadth of project delivery planning issues which may arise and how these have been dealt with. Within these case studies various formats for Project Delivery Plan documents are presented based on the needs of the specific project or programme.

The blank template for a sample Project Delivery Plan is contained on the IChemE website within a protected area. Readers will receive a password with each copy of the book allowing them to access to the template. The actual format of the template cannot be changed but the tool can be used electronically by the reader to fill in the project data as required by the template.

And remember . . .

There will always be someone on your project who is in a great rush to 'start the project' (meaning delivery!) whilst you are pulling together the project delivery plan. The hardest job of the Project Manager is harnessing this energy in the right direction.

➤ Time spent **planning before delivery** is more than compensated by a controlled project delivery.

➤ Start your **delivery in haste and repent in leisure** with the mountain of issues which prolong the project life and reduce the chances of success.

Contents

This book develops the project delivery planning concepts originally outlined in *Project Management Toolkit* (Melton, 2007).

Following approval of a project, the business, the sponsor and the allocated Project Manager are all eager to begin delivery. All too often there is nothing to stop this natural tendency and so project delivery commences with little or no planning. The result is that all stakeholders end up with a less efficient and effective project delivery and outcome for the business.

The role of a Project Manager at this early stage in a project is to 'put the brakes on', engage with appropriate stakeholders and perform some robust and value-added planning, a fundamental part of the project lifecycle.

The project lifecycle

As outlined in *Project Management Toolkit* (Melton, 2007), a project goes through four distinct 'value-added' stages from its start point to its end point (Figure 1-1). Each stage has its own start and end point and each has a specific target to achieve. Effectively each stage can be considered a 'project' within a project.

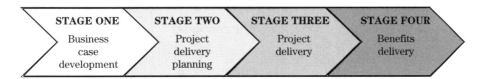

Figure 1-1 The four 'value-added' project stages

Stage One: Business case development

The project start point is usually an idea within the business, for example an identified need, a change to the status quo or a business requirement for survival. At this stage the project management processes should be challenging whether this is the 'right' project to be progressing.

Stage Two: Project delivery planning

This stage is all about planning and the project management processes are used to determine how to deliver the project 'right'.

Stage Three: Project delivery

Effective delivery is all about the control and management of uncertainty. This stage is therefore focused on the controlled delivery – to deliver the project 'right'.

Stage Four: Benefits delivery

The final stage involves integrating the project into the business – allowing the project to become a part of the normal business process, business as usual (BAU).

This book is concerned with Stage Two: project delivery planning, where the start point is typically approval to develop a plan (in readiness for delivery) and the end point is an approved project delivery plan (PDP).

Aims

The aim of this book is to introduce the importance of project delivery planning to an audience of Project Managers who have had both good and not so good experiences when planning and delivering their projects. It provides the reader with education, tools and the confidence to plan projects so that the chances of success are increased.

Figure 1-2 shows an input–process–output (IPO) diagram for this book.

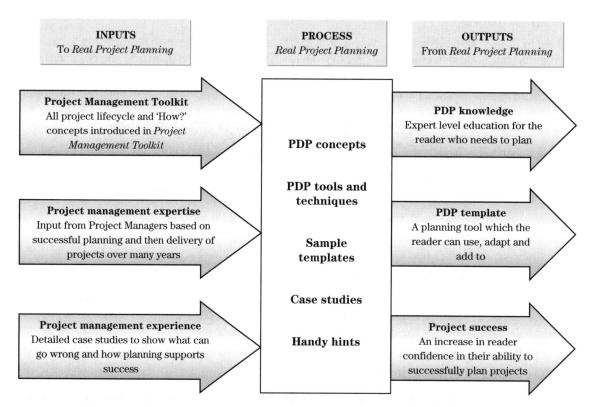

Figure 1-2 The input–process–output (IPO) for this book

Figure 1-2 represents the process by which the aims are to be achieved:

- *Inputs* – lists the inputs to the development of this book.
- *Process* – summarizes the contents of this book.
- *Outputs* – lists the outputs from this book from the perspective of the reader.

Although there are many 'basic' planning tools and techniques available, the aim of this book is to introduce methodologies and principles to support the planning of more complex projects and programmes. Initially though, it will reinforce the project delivery planning tools and concepts introduced in *Project Management Toolkit* (Melton, 2007):

- Asking 'how?' is the second value-added stage in a project and the start of the project delivery planning process.
- Planning a project involves more than generating a time-based activity plan (schedule).
- Planning is all about increasing the chances of success in an uncertain world.
- The process of project delivery planning links Stage One to Stage Three; it aims to plan how to achieve the approved business case.

The book will continue to develop generic tools and techniques which can be applied within any type of organization and any type of project. This will be demonstrated in Chapters 6–8 through the use of a variety of different case study projects.

What is project delivery planning?

It is clear from many project experiences that there is a variable understanding of 'planning' and as highlighted in *Project Management Toolkit* (Melton, 2007), there is common misconception that project planning is merely the time-based plan of activities to be completed. '*When?*' is only one of a number of questions that a project plan must answer if project delivery is to have the best chance of success. Other key questions are:

- **What?:** Is there a clear understanding of what the project is aiming to achieve?
- **What if?:** What are the potential scenarios which could impact project success?
- **Who?:** Does the organization have the resources (capacity and capability) to complete the project; if not, does it have access to appropriate external resources?
- **How?:** There are many ways in which a project goal can be achieved, so a PDP must articulate the actual delivery methodology.
- **How much?:** After 'time', the next area which organizations need to address relates to the amount of investment they should be making and the type of investment (capital or revenue funding, people, equipment, assets and so on).

However, the first question a PDP must answer is '*Why?*' Although a business case will have been developed, it remains the responsibility of the Project Manager to develop and deliver a plan which meets the business requirements, and therefore ultimately deliver something which answers the 'why?' question. A Project Manager can learn a lot about how he should deliver a specific project by asking 'why?'

Short case study

An expanding business required an office refurbishment project because there was not sufficient space in their building to house their employees (current and future forecast). A business case had determined the feasibility of different options to address this business requirement:

- **Option 1:** New office building – determined as too costly for the company at that time.

➤ **Option 2:** New employee working arrangements – increased virtual and home working does not support the required level of customer and colleague contact needed.

➤ **Option 3:** Refurbish existing building – the office utilization (people/m^2) is currently low compared with the national benchmark so this is determined to be the most appropriate solution.

A Project Manager was asked to deliver Option 3 for a specific cost and within a specified time. However, before planning commenced the Project Manager still needed to understand the 'optioneering' so that he could plan appropriately:

➤ Rejection of Option 1 indicates that low cost is a high priority for the company. The increased number of employees signifies an expansion of business but one which is tentative at this stage.

➤ Rejection of Option 2 indicates that the current company culture does not consider that new working practices are required (just allow more people to work in the same space doing the same things). This may signify a resistance by the company management to consider new working practices which are inevitable in a project such as this. It is an indication of how 'soft' issues can impact a project.

➤ Acceptance of Option 3 shows that the company recognizes that some aspects of its working practices are not externally competitive, however the example quoted simply reflects 'hard' issues.

Based on this the Project Manager was able to develop a delivery methodology which incorporated greater elements of business change planning. He did not just accept that he was refurbishing an office; he integrated best practice office 'ways of working' and designs into the project scope. As a result the business not only achieved, but exceeded, its goal by being able to increase output from the business through increased employee capacity **and** effectiveness.

Integration of 'soft' and 'hard' elements

What the above short case study demonstrates is that all aspects of a project need to be planned in order to increase the chances of success: both 'soft' and 'hard' elements. Often the former is ignored and the latter completed at a very tactical level only:

➤ **Soft:** Generally these refer to people, behaviours, relationships and intangible parts of the project or business case.

➤ **Hard:** Generally these refer to the more tangible elements in a project: scope, cost, time, project deliverables and financial benefits.

The effective integration of both elements is what makes a plan a good plan.

Project management is the management of uncertainty and the Project Manager's goal is to progressively increase the certainty of outcome – hopefully a successful outcome. Planning is a crucial phase in this process. There is extensive data available which highlights the value of good planning. It links in to general data on the front end loading (FEL) of the project process. FEL is based on the requirement for organizations to manage limited resources and conflicting priorities:

➤ Organizations need to be sure that they have selected the 'right' project based on robust early definition work.

➤ Organizations need to know that a project will be delivered successfully with some level of certainty.

Good planning is therefore a part of FEL and time spent in the planning stage will deliver benefits that will be seen in remaining project stages and overall project outcomes.

The target for Stage Two is to complete all aspects of the project delivery planning process which:

- Determines the most appropriate way to deliver the project scope to meet the business needs.
- Ensures that the expected outcome is delivered for the business.
- Ensures that the business plans to receive the project and to deliver the benefits.
- Looks at the breadth of options available and the choices to be made to increase the probability of success.
- Highlights the analyses which need to be performed so that the choices are applicable for the specific situation.

Why a PDP is needed

To understand why a PDP is needed we have to understand why good project management practices are needed: to prevent chaos at any stage in a project's lifecycle.

Project chaos is often described as 'utter confusion' and the symptoms we typically see are:

- Projects delivered late or outside of their agreed budgets.
- Projects which don't deliver to agreed quality, quantity and functionality criterion.
- Projects which don't meet the intended business needs.

It is easy to react to the variety of symptoms but such a reaction can lead to further issues. In order to develop sustainable and robust project management practices the root cause of any symptom needs to be found and resolved (Figure 1-3).

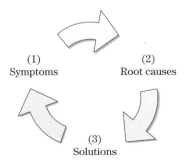

Figure 1-3 Symptoms and root causes

There are many techniques which can be used to identify root causes. The one used here is 'five whys':

- Ask 'why?' a maximum of five times.
- With each 'why?' the cause becomes more specific and therefore actionable.
- Usually the first and second 'why?' will generate further symptoms.
- Usually the third or fourth 'why?' will generate the cause of the specific project issue.
- Typically the fifth 'why?' will generate the root cause which requires resolution at the organizational level.

Within this chapter (pages 7 and 8) the 'five whys' technique has been used to identify project management practices within Stage Two of a project that need to be used to deliver project success: the delivery of sustainable business benefits.

Chaos in Stage Two: project delivery planning

Examples of typical symptoms of project chaos in Stage Two are shown in Figure 1-4. The majority of these are only seen during actual delivery, leading to the conclusion that chaotic or poor project delivery planning impacts delivery effectiveness.

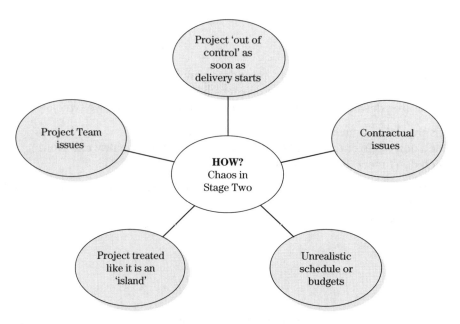

Figure 1-4 Symptoms of project chaos in Stage Two 'how?'

It is therefore useful to review key activities within the planning stage which define 'how?' a project is to be delivered, and what can happen when these are not robustly performed. Typically these are:

- Scope definition as a connected work breakdown structure showing time and resource dependencies and linked to critical acceptance criteria (quality, quantity and/or functionality).
- Risk assessment and management through development of appropriate mitigation and contingency plans.
- Contract definition and management through the development of appropriate contract plans detailing how organizations work together to deliver the project successfully.
- Stakeholder analysis and management through development of engagement and communication plans.
- Collation of the above and all other planning activities into a succinct 'deliverable' – a PDP.

One example of an issue seen during project delivery is a project becoming 'out of control' as soon as delivery starts. This is demonstrated by the following short case study.

Short case study

A project to deliver a new research facility significantly overran on time and cost and at close-out had many contractual disputes and a Project Team in chaos. The project had effectively been 'out of control' since it started and had gone from one crisis to the next, until it finally was completed having failed to achieve many of the critical project objectives:

➤ The business needed the facility to be ready for beneficial use so that newly formed and/or recruited research teams could commence critical research work.
➤ The business case relied on both a maximum capital expenditure and a maximum revenue running cost per annum.

The highly demotivated team conducted a close-out review, including some 'five whys' analyses (Figure 1-5) in order to determine why they were in this situation.

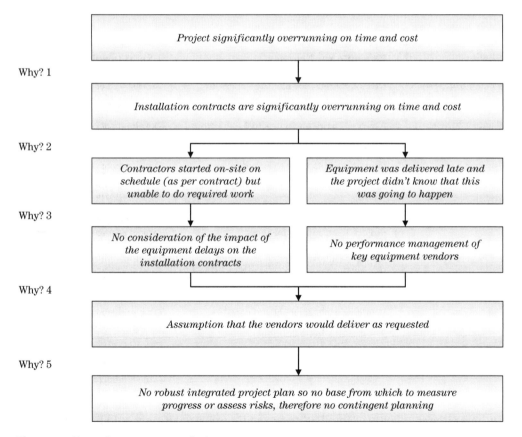

Figure 1-5 Example root cause analysis

In this case the root cause for the lack of project success was the lack of a robust integrated PDP:

➤ The schedule logic wasn't accurate and so was unable to highlight issues when delays started to occur.
➤ The procurement strategy didn't include robust performance management of critical equipment vendors.

➤ A lack of risk assessment meant that no mitigation or contingency plans were developed to support potential and actual problem resolution.

➤ The Project Team were highly segregated and didn't understand the links between the different parts of the project.

After the review, the Project Manager and team were able to better understand what had happened and recommended changes to company procedures so that better planning was incorporated into their project process.

Generic planning issues

Table 1-1 shows the results of more generic 'five whys' analyses which equally point to an issue in Stage Two as the root cause for lack of project success. The common theme throughout all the solutions is the concept of robust project delivery planning – the theme of this book.

Table 1-1 Example root cause analyses – when 'how?' isn't robust

Typical symptoms	Example root causes	Example solutions
➤ Project 'out of control' as soon as delivery starts	Lack of contingency plans to cope with project issues. Lack of risk management	Assessment of risks so that **contingency plans** can be developed
➤ Contractual issues during and after the project	A lack of clarity on contractual obligations	Development of a **contract plan** for each external supplier
➤ Unrealistic schedule or targets	Poor project planning to integrate cost, time and scope issues and link to the business	Development of a **project delivery plan (PDP)** which describes how the project can be delivered
➤ Project treated like it's an 'island'	Lack of communication between the Project Team and the business	Management of **stakeholders** so that the project remains linked to the business
➤ Unmotivated Project Team	A lack of understanding of individual Project Team member responsibilities and their link to the overall project goal	Development of an internal **team communication plan**, **RACI charts** and **document matrices**
➤ Required scope is not delivered	Poor scope management and control	Development of a **scope basis** and a **change control philosophy**
➤ Scope delivered does not enable the required business benefits	No/poor understanding of the benefits which the project needs to enable	Development of a **project delivery plan (PDP)** which describes how the project links to the business

These solutions either explicitly or implicitly suggest that the symptom would be eliminated through the development of a PDP. This 'deliverable' is needed to ensure that:

➤ The delivery of a project is planned to meet a set of specific business requirements.

➤ The project launch is as effective as possible: considering both 'soft' and 'hard' issues.

➤ The project remains 'out of chaos' and therefore 'in control'.

So to answer the general question: 'why is a PDP needed?' the succinct answer is 'to ensure that it has the highest potential of achieving its specified outcome; the highest chance of project success'.

The planning hierarchy

There is a clear vision of success for Stage Two of a project (Figure 1-6) and an associated path of critical success factors (CSFs).

CSF 1 **Approved business case** A robust and articulated business case which describes the business context for the project so that the Project Manager can choose appropriate delivery strategies	CSF 2 **Engaged stakeholders** Ensure that the planning process appropriately involves and engages the sponsor, customer, Project Team and user groups as necessary to build a robust plan	CSF 3 **Capable Project Manager** An appropriately experienced Project Manager with proven skills, knowledge and behaviours as required for the project
CSF 4 **Business plan** Development of the plan to effectively link the project to the business	CSF 5 **Set-up plan** Development of the plan to effectively set-up and launch the project	CSF 6 **Control plan** Development of the plan to effectively control the project

Vision of success
The development of a PROJECT DELIVERY PLAN (PDP) which describes the most appropriate methodology to deliver a project to increase the potential for success through management of uncertainty

Figure 1-6 project delivery planning success

CSF 1: Approved business case

This CSF is the start point for project delivery planning as it sets the business context for all subsequent planning decisions. It is a prerequisite for the development of business, set-up and control plans (CSFs 4–6). The business case should be challenged by the Project Manager so that a real understanding of the reason why this project is needed develops. One method to complete this challenge would be by using the 'Why?' Checklist tool (Appendix 9-1) introduced in *Project Management Toolkit* (Melton, 2007).

CSF 2: Engaged stakeholders

This CSF recognizes that planning is a team activity requiring inputs from a variety of sources. Planning is an opportunity to engage with a broad stakeholder base, often providing people with the first real information on the project and its potential impact on them and/or their business units. In doing so the Project Manager is building an extended team, recognizing it is an opportunity to gain information/feedback from the stakeholder group. Effective two-way communication is an indication of stakeholder engagement.

CSF 3: Capable Project Manager

This CSF recognizes that success at any stage in a project not only needs good organizational support (infrastructure and processes), a good Project Team and a sound basis for the project; it needs a capable Project Manager. This CSF is covered in some detail in Chapter 3, when the overall project organization set-up is discussed.

During Stage Two the Project Manager is responsible for developing the PDP, communicating with all stakeholders and achieving planning success.

CSFs 4–6: Business, set-up and control plans

In terms of project delivery planning, CSFs 4–6 can be considered together within a planning hierarchy (Figure 1-7). This hierarchy proposes three levels of planning which are critical to project success, each linked to a separate CSF.

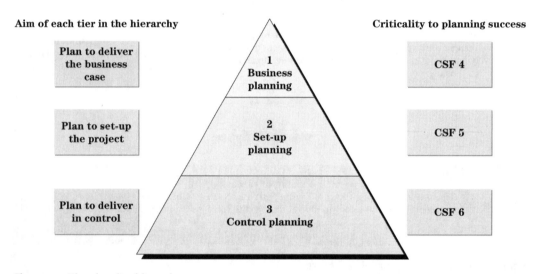

Figure 1-7 The planning hierarchy

Earlier in this chapter the reasons why planning is needed was highlighted and these are all related to the prevention of project chaos and therefore project failure. Planning is necessary because the Project Manager and the business want to:

- Determine how to deliver the business case – *business planning (CSF 4)*.
- Know what should be delivered, who is involved and in what way – *set-up planning (CSF 5)*.
- Know when to deliver and the strategies for controlling this within predetermined cost, quality, quantity and functionality criterion – *control planning (CSF 6)*.

The planning hierarchy (Figure 1-7) is therefore at the heart of the concept of successful project delivery planning and encompasses these three levels so that the certainty of a successful outcome is increased and project chaos eliminated:

- **Business planning** – the effective link between the project and the business.
- **Set-up planning** – the effective administration and launch of the project.
- **Control planning** – the effective link to a controlled project delivery.

Note that project delivery planning can start at any level in the planning hierarchy – the importance is in ensuring that all levels are 'visited' a number of times. However, to confirm the robustness of the completed plan this approach proposes that each tier should be finally checked in sequence, as each depends on the other to some extent, that is:

➤ A good business case (from project Stage One – Figure 1-1) is needed to develop a robust *business plan*.
➤ A good *business plan* is needed to develop an appropriate *set-up plan*.
➤ A good *set-up plan* is needed to develop an effective *control plan*.

The project delivery plan

Project delivery planning is a method to maximize success by planning to manage all areas of uncertainty (Figure 1-8). Each area of uncertainty needs to be managed and these 11 planning themes were originally introduced in *Project Management Toolkit* (Melton, 2007) by use of the 'How?' Checklist (Appendix 9-2), and the basic tools used address them. The 11 themes represent the main components of a PDP.

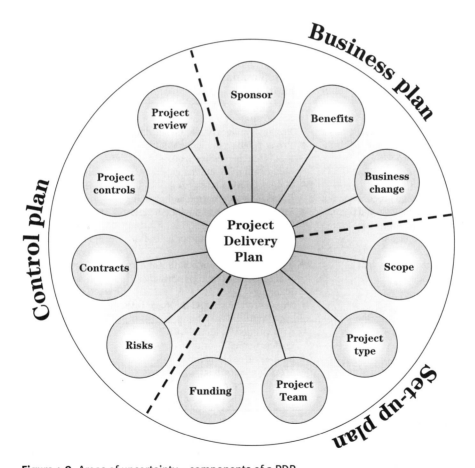

Figure 1-8 Areas of uncertainty – components of a PDP

Although a preliminary PDP will have been generated to support the development of the business case, and therefore pass the 'project approval' stage gate typical in most organizations, Stage Two is concerned with the development of a detailed plan. Every project requires a robust PDP if it is to be successfully delivered and the benefits realized sustainably:

➤ Project delivery planning is the link between what we have committed to delivering (typically in a business case document of some sort) and what we actually do deliver (the completed project).

The formal PDP developed in Stage Two of a project would typically be developed **prior** to any actual project delivery, as it is considered to be:

➤ A formal agreement between the project sponsor and the Project Manager – effectively linking the project to the business.
➤ A Communication and Management Tool for the Project Manager and Project Team – effectively setting up the internal 'workings' of the project.
➤ A communication tool between the Project Manager/Project Team and the external project stakeholders – further integrating the project with the organization in which it sits.

In this book the eleven planning themes have been divided (by chapter) into the three tiers of planning (Figures 1-7 and 1-8) and within each tier additional concepts and tools will be introduced (Table 1-2):

Table 1-2 Summary of planning concepts and tools

	Business planning (Chapter 2)	Set-up planning (Chapter 3)	Control planning (Chapter 4)
Concepts	➤ Business strategy ➤ Sponsorship – selection and role ➤ Customer management ➤ Consultancy process ➤ Benefits management ➤ Sustainability planning ➤ Business change process and plan ➤ Stakeholder analysis and planning ➤ Communications strategy ➤ Benefits mapping and specification ➤ Environment assessment and change readiness	➤ Set-up strategy ➤ Project Manager and team selection ➤ Project management capability model and capability profiles ➤ Value definition and management ➤ Scope quality, quantity and functionality definition and planning ➤ Project and programme organization ➤ Funding strategy ➤ Team skills matrix and organization structures ➤ Team start-up (team building and team processes) and performance planning ➤ Project roadmaps and the stage gate approach ➤ Critical success factors (CSFs) and critical to quality criteria (CTQ) ➤ Scope risk assessment ➤ Work breakdown structure and activity mapping ➤ Business satisfaction analysis	➤ Control strategy ➤ Forecasting ➤ Quality control ➤ Contract and supplier strategy and planning ➤ Risk management process and tools ➤ Cost estimation, planning and contingency ➤ Facilitation process and modes ➤ Cost and schedule risk assessment ➤ Schedule estimation, planning and contingency ➤ Project progress measurement and performance management ➤ Critical path of risks ➤ FMEA, risk flowcharts and checklists
Tools	➤ Stakeholder contracting checklist ➤ Sponsor Contract Planning tool ➤ Benefits realization plan ➤ Sustainability plan ➤ Project charter ➤ Communications Planning tool	➤ Project Manager selection checklist ➤ Team selection matrix ➤ Project Team role profile ➤ Team start-up checklist ➤ Roadmap decision matrix ➤ Finance strategy checklist ➤ Activity plan ➤ Scope definition checklist	➤ Project Scenario Tool ➤ SWOT table ➤ Critical path of risks table ➤ Risk management strategy checklist ➤ Contract plan ➤ Supplier selection marix

However, the concept of challenging a completed PDP through using the 'How?' Checklist remains valid and a reinforcing concept throughout each of the case studies (Chapters 6–8).

The remainder of the book is structured around the concept of the planning hierarchy (Figure 1-7) and the 11 planning themes (Figure 1-8). The flow of the chapters is put together so that the reader can also commence project delivery planning in a number of ways (Figure 1-9). What is important is that the final PDP cannot leave the 'planning cycle' until a number of iterations have been completed.

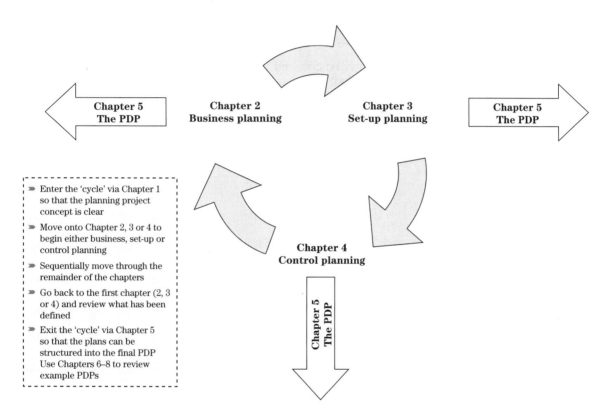

Figure 1-9 The project delivery planning cycle

In this way a PDP can be developed which:

- Links to the business and the measurable business benefits.
- Contains a robust plan for the delivery of the scope which itself is linked to the delivery of the benefits.

This will increase the chances of success in terms of an organization delivering the 'right' project, and in delivering that project 'right'.

As well as introducing concepts and tools, short case studies are used in each chapter to further demonstrate key aspects of business, set-up and control planning.

And remember . . .

- There are four value-added stages in the project lifecycle.

- Stage Two is all about the development of a PDP.

- Project delivery planning should deliver a robust plan which mitigates typical causes of project chaos often seen in Stage Two.

- At the end of Stage Two it should be clear how a project will be delivered and how that delivery plan will support the realization of a set of specific business benefits.

- Stage Two supports an organization in setting up a project for successful delivery.

- The conclusion of Stage Two can always be checked via use of the 'How?' Checklist introduced in *Project Management Toolkit* (Melton, 2007).

2 Business planning

In the context of project delivery planning, business planning is the way that we link the project to the business. It is based on the approved project business case which explains 'why' the project is needed in the context of the organization. Within the project delivery plan (PDP) there needs to be a further explanation of 'how' specific elements will be delivered.

Historically this has been the weakest part of a PDP. This is because it relies on the Project Manager being 'outward looking', towards the organization, as opposed to 'inward looking', towards the project. However, as described in Figure 1-6, there are two planning Critical Success Factors (CSFs) which feed into the development of a business plan (which is a CSF itself):

➡ *CSF 1*: Approved business case – the 'hard' starting point for the project.
➡ *CSF 2*: Engaged stakeholders – the 'soft' starting point for the Project Manager.

The physical link between the project and the business is usually through the relationship between the Project Manager and the project sponsor, but also through the project sponsor and the customer. It is these relationships which need to be planned:

➡ The Project Manager develops a plan to deliver the agreed scope.
➡ The sponsor ensures that the plan will enable the business need to be met but also works with the customer to understand how the business will be ready for the project.
➡ The customer understands how the project will deliver the required scope, as well as what else needs to change within the business in order for the project to enable the delivery of the business benefits.

Additionally the business plan must provide a robust methodology for the realization of the business benefits, the delivery of which prove that the business case has been met. All projects need:

➡ A Project Manager with a clear understanding of the project goal – a successful outcome.
➡ A business issue to solve and a sponsor to own both that issue and a successful resolution.
➡ A customer and end user group who will eventually 'own' and integrate the results of the project into their day-to-day operations.

Projects can fail when these essential roles, relationships and strategies are missing or are not effectively managed. Developing a business plan is the start of this relationship management:

➡ Planning to manage the sponsor, customer and senior stakeholders.
➡ Planning to deliver a change to the business that is value add to that business.

What is a project business plan?

Within a PDP, a project business plan is the formal articulation of 'how' the project will deliver the business case. It links the project to the business and covers the following three planning themes:

➣ Sponsorship.
➣ Benefits management.
➣ Business change management.

How to develop a project business plan?

It is not unusual for a Project Manager to develop the majority of this type of plan before launching the project and before the full Project Team is on board. It is important that the Project Manager begins relationship management as soon as he is assigned a project.

To develop an effective business plan a Project Manager needs to build relationships, communicate effectively and piece together the detailed benefits realization plan which will meet the approved business case. To do this a Project Manager needs to behave as a consultant and use consultancy tools to facilitate key relationship building, so that all aspects of the business plan are developed (Cockman et al., 1999).

The consultancy process

A key relationship which needs to be developed and managed is the one between the Project Manager and the project sponsor. Figure 2-1 shows a typical consultancy lifecycle model which is appropriate to consider for the development and management of this relationship.

A Project Manager will go through each phase of the consultancy lifecycle with all stakeholders including the sponsor, the customer and the Project Team.

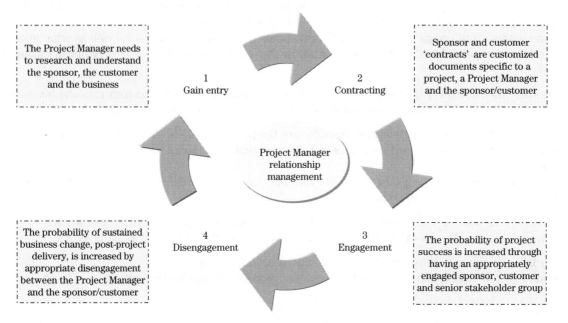

Figure 2-1 The consultancy lifecycle and project business planning

Step 1: Gaining entry

The first stage to building a relationship is to make initial contact. Gaining entry is the process of making an effective initial contact and establishing early expectations. Gaining entry is important as it can:

- Enable the stakeholder to clearly state the need.
- Get some idea of the readiness for the project and any associated business change.
- Establish your credibility.
- Start/develop a relationship on which to build.
- Establish common ground.
- Find out how welcome/unwelcome you are.
- Get some idea of how the project will impact the specific stakeholder and associated business processes.
- Get some idea of the size of the problem being solved by the project or the business challenge being faced.
- Find out how urgent the project is.
- Understand the impact of any existing relationships as they may help or hinder effective relationship building between the Project Manager and the stakeholder group.

Step 2: Contracting

The second stage in building an effective relationship is to agree a 'contract'. A contract is a way of making expectations explicit. A good 'contract':

- Defines the overall expectations of the relationship.
- Clarifies specific actions or activities which are to be performed.
- Provides the parameters and freedom to act within them.
- Sets agreed goals and objectives.
- Establishes the ground rules for behaviour.

Step 3: Engagement

After an effective relationship has been established, the process of engagement is concerned with the ongoing management and development of the relationship. Engagement means:

- Management of organizational politics.
- Ongoing commitment to the agreed 'contract' from both parties: (Project Manager and stakeholder).
- Assessment of 'contract' performance through openness and trust.
- Being clear about the future vision so that you know when it has been achieved.

Step 4: Disengagement

Upon realization of the business benefits, a Project Manager needs to disengage from a 'contract'. Disengagement means:

- Deciding that the 'contract' has been delivered and there is no ongoing need for the relationship as related to the project.
- Closing out the relationship: agreeing on what has been delivered, sustainability, and the success of the 'contract'.
- Letting go and saying goodbye!

Tool: Stakeholder Contracting Checklist

A tool has been developed based on the consultancy lifecycle a tool (Figure 2-1), to support effective planning of all stakeholder contracts in a project environment (Table 2-1):

Table 2-1 The Stakeholder Contracting Checklist explained

Planning Toolkit – Stakeholder Contracting Checklist			
Project:	<insert project title>	**Project Manager:**	<insert name>
Date:	<insert date>	**Page:**	1 of 2
Contract set-up			
Stakeholder name <insert name> **Link to this project** <insert formal link to the project and any formal role on it. Add any informal influence the stakeholder has on the project or on other stakeholders>			
Stakeholder background			
Likely stakeholder engagement on this project <insert comments on how likely it is that the stakeholder will support the project and if there is any information on current views> **Likely stakeholder knowledge of this project** <insert what knowledge the stakeholder may have of the project prior to formal relationship building with the Project Manager>			
Project requirements from stakeholder			
List any information the project will need from this stakeholder <it is usual that a Project Manager might expect senior stakeholders to 'open doors' for himself or the team. Insert a list of these types of needs as well as any data or documentation> **List any decisions the project will need from this stakeholder** <insert critical decisions and likely timing. This may include decisions on release of critical resources: people, assets and/or funding> **List likely impact this stakeholder may have on the project outcome** <insert how this stakeholder could impact the project: positively or negatively>			

(continued)

Table 2-1 (Continued)

Planning Toolkit – Stakeholder Contracting Checklist			
Project:	*<insert project title>*	**Project Manager:**	*<insert name>*
Date:	*<insert date>*	**Page:**	2 of 2
Stakeholder requirements from project			
List any information this stakeholder will need from the project *<insert likely project data which will be of interest to this stakeholder. Bear in mind that the stakeholder may not be able to articulate this at the beginning of the relationship>* **List likely impact this project may have on the stakeholder business operations** *<insert how this project may change the stakeholders business at any stage in its value chain>* **List any current stakeholder concerns** *<insert any concerns – whether articulated or assumed so that a plan to deal with them can be developed>*			
Contract agreement			
Outline what level of communication has been agreed (throughout the project) *<insert agreed arrangements for communicating: for example verbally, in writing and also the frequency>* **Confirm that the contract (decision and information flow) has been agreed between the Project Manager and the stakeholder** *<insert whether you have an agreed explicit contract or the stage you are at in the consultancy lifecycle>*			

The contracting checklist is intended to be used internally by the Project Manager and completed 'off-line'. It is not suggested or advised that this is used in a contracting situation face-to-face with a stakeholder as but, a method to collect and collate confidential data which supports development of appropriate plans.

Contract set-up

A Project Manager should first develop a relationship with the project sponsor. Then, based on stakeholder identification followed by categorization (Figure 2-15, page 53), develop other appropriate relationships such as with the customer.

Stakeholder background

Everyone has a history, and it is important that a Project Manager research the history of the stakeholder as it impacts the project.

- Occasionally a Project Manager may be familiar with the stakeholder, through a previous project or business encounter.
- Normally a sponsor has a reputation within the organization based on previous management behaviours.
- Typically the stakeholder will know something of the project and therefore have some perspective.

Project requirements from stakeholder

A stakeholder is needed for many reasons, and the role of the Project Manager is to define these needs and secure agreement on them (Table 2-2).

Table 2-2 Project needs checklist

Project need	Stakeholder role	Example
1. Project team members	➤ Release of appropriate people to support the project	➤ A Quality Manager may need to make some of his laboratory analysts available to support design within a Laboratory Refurbishment Project
2. Tangible assets	➤ Make assets available at the time needed by the project	➤ A Production Manager needs to handover the manufacturing facility to the project so that agreed modifications can be made
3. Approvals	➤ Make decisions at various points (such as stage) gates to move the project forward or key document approvals	➤ A Business Manager needs to approve the business case for a project before any funding will be released and the project formally launched
4. Data	➤ Allow access to data during the project and ensure the data is accurate	➤ A Business Manager needs to update the team on sales forecasts so that the manufacturing capacity sensitivity analysis can be conducted at key decision points

A Project Manager should conduct a risk assessment to fully understand the impact on the project if the stakeholder does not deliver as required. Occasionally the 'contract' is not successfully agreed and the impact of this also needs to be assessed.

For example, prior to the start of a project to install a new laboratory HVAC system the Laboratory Manager would not agree to the requested date to make the laboratory available (and therefore shut down all testing). Although the testing that week was important it was no more critical than any other week. However, without the new system the internal environment did not meet legal requirements in terms of protecting laboratory personnel. In this case the Safety Officer (another project stakeholder) confirmed that the laboratory would be closed down if the HVAC wasn't installed as soon as possible.

Stakeholder requirements from project

On any project the stakeholder needs will depend on the specific stakeholder role and category (page 53). Stakeholders may need information for a variety of reasons and these may or may not have an impact on the project itself (Table 2-3).

Table 2-3 Stakeholder information needs

Stakeholder information need	Potential project impact
1. To make a project related decision	Delayed decision making can stop a project 'flowing' and so all relevant information needs to be issued to the stakeholder in advance of the decision
2. To make a business decision	A delay to a business decision could have a much broader impact although it may be minimal on the project
3. To support effective communication in the business	Engaging stakeholders as project communicators is an effective method to disseminate clear, aligned project information
	It can help to manage all types of stakeholder and pre-empt any unnecessary external interruptions or diversions
4. To know what is going on	Stakeholders who want information for no other purpose than so they know what is going on, can impact a project if not well managed

This type of data will support the eventual stakeholder management plan (page 52) as well as being prepared during contracting and engagement. Often a Project Manager will be meeting the stakeholder needs through a route other than that requested by the stakeholder as a strategy to 'protect' the project from the stakeholder.

For example, an engineering manager has indicated that he wants to approve all engineering specifications on the project as his method of ensuring that the completed facility will be safe and meet all legal requirements. The Project Manager knows that this will cause many delays as the engineering manager will effectively become the 'bottleneck' in the project. To remove the issue the eventual solution is:

- Ask the engineering manager to be a part of the team selection process (whether internal or external) so that he can ensure that all team members involved in engineering specification are appropriately qualified to do so.
- Ensure that the engineering manager approves the engineering quality plan. This will be the document which defines the codes and regulations under which the engineering will proceed and which also includes project specific document matrices showing who will generate, review and approve engineering specifications (not him).
- Invite the engineering manager to one of the design quality audits.

In this way the engineering manager got the outcome he wanted and the Project Manager maintained control of the project.

Contract agreement

This tool aims to support the research process that will help a Project Manager to 'gain entry' effectively and then develop a 'contract' which is beneficial to the project. There are other associated tools (page 27) which will support actual contract development and the eventual formal output which is seen by a stakeholder (page 30).

Sponsorship

Many organizations misunderstand sponsorship and therefore do not reinforce the importance of having an appropriate and active sponsor on a project. At an organizational level there are a number of reasons why project sponsorship is needed:

➤ Organizations change, business strategy changes and the external environment changes. A project can be viewed as an unchanging constraint in an organization, as an 'island'. This is unacceptable, therefore there needs to be some controlled way of linking the project to the business.

➤ There needs to be clear organizational accountability for the expenditure of any resource which is not a part of 'business as usual' (BAU) and not managed within 'normal' business processes.

Business strategy

During the development of the PDP the Project Manager will need access to appropriate and accurate data to make the 'live' link between the project and the business. Typical data needed during the planning stage is:

➤ *Organizational position*: Where in the organization will this project sit when it is completed? Which business unit will integrate the completed project into their BAU?

➤ *Business vision/strategy*: What is the overall goal for the business unit into which the project will deliver some measurable benefit? How is the business hoping to achieve this goal and how will it measure success?

The project operates in an organizational environment and the process of sponsorship should ensure that changes in this environment are assessed for any potential impact on the project. A typical business impact assessment process is shown in Figure 2-2.

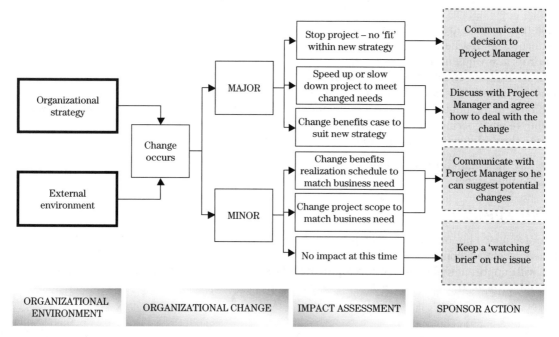

Figure 2-2 Business impact assessment process

The sponsor effectively acts as a filter ensuring that only those organizational changes that have an impact on the project are brought to the attention of the Project Manager. The Project Manager and sponsor need to agree such a process as a part of their contract (page 30).

In developing a quick overview of the business strategy within the business plan section of the PDP the Project Manager gains an understanding of the organizational culture. This is important for other areas of the plan such as business change management (page 47).

Sponsor selection

One of the main reasons why sponsorship is not successful relates to sponsor selection. In order to support the project appropriately the sponsor needs to be at the lowest possible level in the organization to have the necessary authority to make decisions (and gain organizational information as needed to progress the project). Common mistakes are:

➤ Having a sponsor who is too senior.
➤ Having a sponsor who has no authority over the business area into which the project is delivering some business change.

Sponsorship should create an infrastructure of commitment for the project and this can only be successfully developed through appropriate sponsor and reinforcing sponsor selection.

Consider two projects in the example organization shown in Figure 2-3:

➤ Project 1 is an upgrade of a functional work area in Division A:
   ▷ The people working in Function 2 will be impacted by the change – they are the targets. The upgrade allows Function 2 to deliver to agreed performance levels.

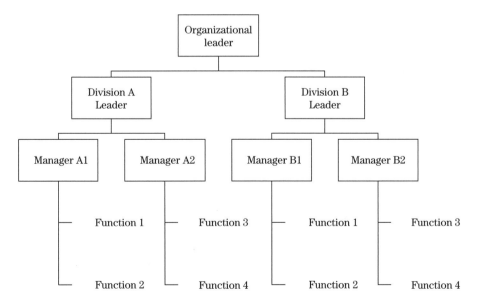

Figure 2-3 Sponsor selection

⊃ Function 2 Manager is the champion for the project – he pushed for the change and supported the development of the business case.

⊃ Manager A1 is the sponsor for the project – he has complete budget authority as well as authority over the function.

⊃ In this scenario it is typical for Division A Leader to be nominated as the sponsor when in fact his authority is not needed.

⧓ Project 2 is the implementation of new HR systems and processes:

⊃ All people in the organization are impacted – they are all targets.

⊃ The Organizational Leader is the sponsor and due to the size of the project he has delegated part of his authority on this project to Division A and B Leaders making them reinforcing sponsors.

⊃ Division A Leader is the champion for the change.

⊃ In this scenario it is typical for the Organizational Leader to attempt to sponsor the project himself without developing a shared commitment for change through reinforcing sponsorship.

The sponsor for a project would therefore be chosen based on position within the organization rather than any personal attitude to the change. This latter aspect needs to be incorporated into the management of the sponsor.

The role of the sponsor

A sponsor is accountable for the realization of the business benefits; for delivering the business change which was approved via the formal business case and for ensuring that the project still fits within the organizational environment.

Although there is a very typical role for a sponsor as distinct from the Project Manager or customer, there still needs to be a formal relationship built to make the role explicit. Table 2-4 shows how the sponsor is involved with typical planning activities.

Table 2-4 The role of the sponsor during planning

Planning activity	Sponsor role	Project Manager role
Stakeholder identification, categorization and planning (page 52)	⧓ Accountable ⧓ Actively involved	⧓ Responsible ⧓ Actively involved ⧓ Develop the formal output – stakeholder management plan (Appendix 9-6)
External communications plan (Table 2-8)	⧓ Accountable ⧓ Give clear direction ⧓ Review and approve	⧓ Responsible ⧓ Develop the formal document based on stakeholder plan and sponsor direction
Project charter development (Table 2-7)	⧓ Accountable ⧓ Give clear direction ⧓ Review and approve	⧓ Responsible ⧓ Develop the formal document based on business case and sponsor direction

(continued)

Table 2-4 (Continued)

Planning activity	Sponsor role	Project Manager role
Benefits realization planning	⮞ Accountable ⮞ Give clear direction ⮞ Review and approve	⮞ Responsible ⮞ Develop the formal output – benefits realization plan (Table 2-11) based on business case and sponsor direction
Business change planning (page 49)	⮞ Accountable ⮞ Agree to a clear set of responsibilities during delivery ⮞ Review and approve	⮞ Responsible ⮞ Develop the formal output – business change plan (Figure 2-12) which should incorporate a sustainability plan (Table 2-13)
Project delivery planning	⮞ Review and approve ⮞ Specific review points would be linked to having the best people, sponsor decisions and ensuring overall cost and timeline matches the business case	⮞ Accountable and responsible ⮞ Develop the formal PDP document (Chapter 5)
Project organization development	⮞ Agree to obtain specific resources based on skills needed to achieve the goal ⮞ Review and approve high-level organization only	⮞ Accountable and responsible ⮞ Develop the project organization (Chapter 3, page 84) ⮞ Develop the project RACI (Appendix 9-9)
Project roadmap development (Figure 3-13, page 106)	⮞ Agree go/no go decision points during project delivery ⮞ Review and approve	⮞ Accountable and responsible ⮞ Develop the formal document based on proposed strategy for project delivery and typical strategies for similar project types
Project cost and schedule development (Chapter 4, page 185)	⮞ Review and approve high-level plan only: cost matches approved funding and schedule matches approved benefits realization plan	⮞ Accountable and responsible ⮞ Develop the formal control documents based on the agreed business case

The sponsor has a major role during the planning stage particularly with respect to the business plan – the plan which links the project to the business.

In order to make the role explicit the consultancy cycle (page 16) should be used by the Project Manager:

⮞ *Gaining entry*: Build an effective and appropriate relationship – get to know each other.
⮞ *Contracting*: Make expectations explicit (two-way) – build a RACI Chart (Appendix 9-9) and write a contract (Figure 2-4).
⮞ *Engagement*: Manage the relationship effectively – use the tools developed during contracting and assess the effectiveness of the relationship. Maintaining a good and beneficial relationship with the sponsor throughout the project is crucial to the delivery of the business plan.

⬤ *Disengagement*: Close out the relationship – review performance against contract and learn from the experience. At the end of the project the Project Manager must understand how he intends to professionally disengage from the relationship with the sponsor. Without planning an appropriate disengagement strategy Project Managers can be associated with the project outcome long after the project scope has been successfully delivered and the Project Team has been disbanded.

A good contract between the Project Manager and the sponsor at the outset will enable expectations to be appropriately set and disengagement parameters agreed, for example:

⬤ When the plant has produced its first kilogram of product.
⬤ When the last contractor leaves site.
⬤ When the last invoice has been paid and the finances can be closed out, etc.

These usually relate back to project success criteria defined at the start of the project and it is important that these link into the effective handover of the project to the end user. The complex relationship between the sponsor, Project Manager and end user/customer needs to be managed to ensure an effective handover and appropriate Project Manager/Project Team disengagement from the project.

The four elements making up the consultancy cycle need to be incorporated into the sponsor contract and into the contract planning process.

Sponsor assessment

During the disengagement stage of a sponsor/Project Manager relationship it is useful to reflect on the success of the relationship so that both can learn from it. The sponsor contract can be used:

⬤ Go through each point in the contract and ask 'did this happen?'
⬤ If the answer is 'yes' then conduct a '5 whys' analysis to ensure that the root cause for the good outcome is understood and can therefore be repeated.
⬤ If the answer is 'no' then conduct a '5 whys' analysis to ensure that the root cause for the negative outcome is understood and therefore not repeated.
Review the level of positive and negative responses, the types of root causes and conclude on the success of the relationship.

At the end of the session the sponsor and Project Manager should use feedback from other team members (project and production team) to confirm the evaluation.

Tool: Sponsor Contract Planning Tool

The sponsor/Project Manager contract is perhaps the most critical stakeholder relationship for a project and as such it is right that a Project Manager spends time on the process of developing the contract.

Apart from using the Stakeholder Contracting Checklist (Table 2-1) to manage the contracting process it is useful to explicitly plan the sponsor contract. Table 2-5 shows how the consultancy cycle can be used as such a planning tool.

Table 2-5 Sponsor Contract Planning Tool explained

Planning Toolkit – Sponsor Contract Planning Tool			
Project: *<insert project title>*		**Project Manager:** *<insert name>*	
Date: *<insert date>*		**Page:** 1 of 2	
Consultancy stage	**Input**	**Process**	**Output**
1. **Gaining entry** Plan to make the most appropriate 'entry' and achieve the aim	➤ *<insert current relationship with the sponsor>* ➤ *<insert any known sponsor views on the project or Project Manager>*	*<choose from: formal introduction by a third party, telecon, formal 1-1, informal 1-1>*	➤ Understand sponsor perspective and potential issues or concerns ➤ Sponsor awareness and understanding of the project ➤ Sponsor ready for contracting
2. **Contracting** Plan this stage to ensure that the contract is complete and covers all areas	➤ *<insert formal and informal data needed to generate the contract>* ➤ *<examples are draft business plan, draft stakeholder plans, draft communications plans>* ➤ *<insert current thoughts on what decisions are needed, and how the contract will be managed>*	*<insert process which would typically be through 1-1 contact>*	➤ Sponsor agreement on project success criteria ➤ Sponsor agreement on stakeholder involvement ➤ Communications agreements ➤ Confirmed business plan ➤ Confirmed contract with Project Manager ➤ Complete agreement and understanding on the two-way relationship through all stages of the project

(continued)

Table 2-5 (Continued)

Planning Toolkit – Sponsor Contract Planning Tool			
Project: <insert project title>		**Project Manager:** <insert name>	
Date: <insert date>		**Page:** 2 of 2	
Consultancy stage	**Input**	**Process**	**Output**
3. **Engagement** Plan this stage and input the plan into the contract	⟫ *<insert formal and informal data needed to manage the contract>* ⟫ *<examples are business plan, sponsor contract, high-level progress, any stakeholder issues, general concerns>* ⟫ *<insert current thoughts on how the effectiveness of the contract is to be measured>*	*<insert agreed process: based on communications plan, review plan and agreed informal 1-1 sessions>*	⟫ Sponsor happy with progress ⟫ Sponsor support in dealing with stakeholder issues ⟫ Sponsor understanding of risks/concerns and appropriate support to deal with them ⟫ An effective and successful working relationship
4. **Disengagement** Plan the disengagement criteria	⟫ *<insert formal and informal data needed to exit from the contract>* ⟫ *<examples are sponsor contract, agreed project success criteria, benefits realization plan>*	*<insert formal process typically a formal close-out session with the sponsor>*	⟫ Sponsor confirmed role with customer/end user ⟫ Project Manager role ended ⟫ All close-out criteria achieved ⟫ Effective learning from the relationship

This tool is an internal document and is not intended to be used explicitly with the sponsor. Developing a good relationship at the start of the project will support project success but needs to be planned. The use of techniques to make this often informal process more overt is common in business change projects but less so in traditional engineering projects, where a Project Manager may be dealing with a complete steering group rather than one individual. Differences in the sponsor role are not unusual as this is typically dependent on organizational history.

Short case study

A Project Manager had been assigned to deliver a production revamp project. He had not worked closely with the project sponsor previously but had heard rumours of other Project Managers having problems with changing goal-posts, unexpected and detailed involvement in project scope decisions and an inability to let go of the project management resource at project handover. To help manage the situation the Project Manager first of all developed a sponsor contract plan (Table 2-6) which he used throughout the project.

Table 2-6 Example of Sponsor Contract Planning Tool

Planning Toolkit – Sponsor Contract Planning Tool			
Project: Production Revamp		**Project Manager:** Fred Jones	
Date: January 2007		**Page:** 1 of 1	
Consultancy stage	**Input**	**Process**	**Output**
1. Gaining entry	➤ No current relationship between sponsor and Project Manager ➤ A common third party who would add credibility to selection of Project Manager	➤ Get a third party introduction from someone credible ➤ Potentially use Production Director who worked with Project Manager on previous project	➤ Sponsor assured that he has the right Project Manager allocated
2. Contracting	➤ History of poor sponsorship and meddling ➤ Approved business case ➤ Anecdotal data on the business situation ➤ Draft and then final PDP	➤ Series of informal and then more formal 1-1 sessions ➤ Start with 'blank paper' and then gradually introduce 'straw men' ➤ Meet frequently as a part of project planning	➤ A one-page contract (Figure 2-4) – agreed by both sponsor and Project Manager ➤ A RACI chart detailing roles during the design phase ➤ A project charter ➤ Agreed resource from the production team
3. Engagement	➤ Sponsor contract ➤ PDP including high-level RACI Chart ➤ Production representative on the design team that sponsor is happy with	➤ Significant face-to-face time especially during planning and then design ➤ One-page bulletins ➤ Check progress of relationship using RACI ➤ Provision of high-level project progress data: schedule, cost plan and risk report	➤ Sponsor abides by the contract ➤ Sponsor delivers as per the RACI chart
4. Disengagement	➤ Sponsor contract ➤ Handover certificate ➤ Project close-out report (detailing success criteria achieved) ➤ Benefits Tracking Report	➤ Informal team close-out celebration following handover ➤ Formal 1-1 with sponsor based around a brief close-out report	➤ Sponsor happy that the working relationship is concluded and the production unit is back on line

The first stage in the sponsor contract plan was to 'gain entry', which the Project Manager successfully did via an introduction from a senior executive who the sponsor was known to value and respect.

The contracting stage took a long time to complete as the sponsor was initially uncomfortable with discussing 'hard' issues such as:

➤ When the project was over.
➤ What documents he would and would not see.
➤ What decisions he would and would not be involved in.

The contracting discussions brought out a lot of the behaviours which the Project Manager had heard about and enabled him to build a much stronger contract (Figure 2-4). This gave the sponsor a clear role and a clear set of activities which he needed to complete.

Production Revamp Project
Sponsor Contract

PROJECT SUCCESS CRITERIA
➤ Deliver the Production Revamp Project on schedule and within the approved budget (as set out in the Business Case and within any approved change control plan)
➤ Ensure that the final facility is capable of producing at the required capacity and quality

PROJECT MANAGER WILL PROVIDE
➤ The PDP for review and approval
➤ High-level design plans for information only (such as agreed facility layout and procurement plan)
➤ Current information on cost, schedule and risks

HANDOVER CRITERIA
➤ All aspects of construction are complete
➤ All project documentation has been handed over
➤ Facility is ready to be put back on-line
➤ Production team have been briefed by the project design team with regard to new areas of functionality

COMMUNICATION
➤ A one-page Sponsor Bulletin will be issued every 2 weeks in the design phase and every week during construction
➤ Meetings as per schedule (no canceling)
➤ Ad hoc phone calls/emails as necessary – both are available to each other as needed to deliver the approved business case

HOW WE WILL WORK TOGETHER
➤ Sponsor and Project Manager will meet each week until the project delivery plan (PDP) is completed and approved. After this point the meetings will revert to every month until construction starts
➤ Sponsor and Project Manager will agree a high-level RACI to confirm exact role and responsibilities during the design phase
➤ Project Manager will notify sponsor of any project issue which is likely to impact achievement of the project success criteria
➤ Sponsor is to allow design decisions to be made by his nominated representative
➤ Sponsor meetings are confidential unless otherwise agreed

SPONSOR WILL PROVIDE
➤ Access to current production team and area for the duration of the project
➤ Ensure that production team understand their role in supporting the design, handover and commissioning phases and provision of appropriate resources to do this
➤ A nominated production representative who has authority to make design decisions
➤ Handover of the facility for construction on schedule
➤ Timely decisions regarding submitted changes which impact cost or schedule adherence
➤ Support of all project decisions made by others
➤ Any updated business information which might have an impact on the project

Project Manager (sign and date)	Sponsor (sign and date)

Figure 2-4 Example sponsor contract

In this instance the time spent developing an appropriate working relationship with the sponsor was well worth it:

- The sponsor focused on a bounded set of responsibilities which enabled the project to progress without inappropriate meddling. This was reinforced through use of a RACI chart which clearly stated what the sponsor should and should not do.
- The sponsor was able to engage with his own team within production which helped him to think about 'life after the project'.

There were some issues, as expected, but these were small in comparison with historical evidence of previous sponsor behaviours. In the end the Project Manager was able to effectively deliver the project and then disengage as the production facility went back on-line.

Tool: The Project Charter

A key document which can be used to support an effective sponsor/Project Manager relationship is the Project Charter (Table 2-7). This document is usually developed as a part of the early stages of sponsor contracting. It allows both sponsor and Project Manager to translate the approved project business case into a succinct, easily understood document as well as adding agreed elements of the PDP (how the project will be delivered).

Table 2-7 The Project Charter Planning Tool explained

Planning Toolkit – **Project Charter**			
Project: <insert project title>		**Project Manager:** <insert name>	
Date: <insert date>		**Page:** 1 of 1	
Project description		**Project delivery**	
Sponsor <insert name of assigned sponsor who is accountable for the realization of the benefits>		**Project Team** <insert the names of the key team members and their role on the project>	
Customer <insert who/what in the organization will use the final project deliverable>		**Additional resources** <insert any additional resources needed to deliver the project outside of the Project Team>	
Project aim <insert the vision of success as seen from the customer's perspective>		**Critical Success Factors** <insert those factors without which the project cannot be a success>	
Project objectives <insert the measures to prove that the project has been delivered>		**Organizational dependencies** <insert any organizational factors which could have an impact on the project and are outside its scope and control>	
Benefits <insert the benefit metrics which prove that the business case has been delivered>		**Risk profile** <insert a summary of the main threats and opportunities in progressing this project and therefore the risk profile>	
Final deliverable <insert the tangible output from the project>		**Critical milestones vs deliverables** <insert a list of critical milestones, each linked to an interim deliverable>	
Interim deliverables <insert the key interim deliverables required to create the final deliverable>		**Project delivery approach** <insert a high-level description of the process being followed to deliver the project>	

A good starting point for a Project Charter is the Simple Benefits Hierarchy (Appendix 9-4) as this has the majority of the project summary information required.

Project description

➤ *Sponsor*: By agreeing to put his name on the project charter the sponsor is once again overtly contracting to deliver the business benefits. As the project charter is the first document likely to be seen by all types of stakeholder (both within and external to the project) it is one way of communicating the main accountability for the project.

➤ *Customer*: The customer, or more likely customer group, is also named on the charter as a method of broad communication. The customer is the part of the organization that requires the output from the project and there will be many stakeholders within the customer group who have an interest in the project and who also may impact the outcome through articulation of their requirements.

➤ *Project aim*: This should be a succinct description of the vision of success for the project as articulated by the customer.

➤ *Project objectives*: The project will need to deliver a scope of the required quality, quantity and functionality within cost and time constraints. These are determined through development of the customer requirements as summarized in the project aim and as approved within the business case.

➤ *Benefits*: The project will need to enable a set of business benefits which prove that the business case has been delivered. There is a relationship between project objectives and business benefits which needs to be understood so that the benefits are realized sustainably.

➤ *Final deliverable*: A project delivers an output and this output is usually some tangible deliverable which has been requested via the customer and approved via the business case.

➤ *Interim deliverables*: These are tangible project outputs which are critical to the delivery of the final deliverable. They are usually linked to the project schedule critical path and/or the CSFs and provide a useful progress measurement basis.

Project delivery

➤ *Project Team*: The core team members who will work closely with the Project Manager to deliver the project should be named and their main role articulated. The aim here is to communicate the appropriateness of their selection in order to add credibility to the project.

➤ *Additional resources*: Recognizing that not all the critical resources will have been selected at this stage, and may not even be from within the organization, this is an opportunity to highlight specific skills, knowledge or capabilities which are required within the team. Additional funding and/or assets may be highlighted here.

➤ *Critical success factors (CSFs)*: It is useful to define quantifiable/measurable and identifiable actions/activities that may impact the overall success of the project and hence achievement of the vision of success.

➤ *Organizational dependencies*: No project can be delivered without some impact or influence on the organization into which it is being delivered. Often the organization itself can impact the project through non-delivery of business changes outside of the scope of the project or through reliance on parallel projects. These should be identified as early as possible.

➤ *Risk profile*: In the development of the business case the main threats and opportunities will have been developed and these need to be summarized so that a high-level understanding of the project risk profile is known by all stakeholders.

➤ *Project delivery approach*: Although the PDP will not be fully developed at this stage, the general approach will have been agreed and scope, cost and time boundaries be based on this.

An effective and agreed Project Charter is an excellent document to kick off consistent communications both within and external to the project. An example of a completed Project Charter is included in the business plan case study at the end of this chapter (page 64).

Communications strategy

One of the main activities of the sponsor is to support the generation of a project specific communications strategy. This strategy is generally concerned with communications external to the project, as it is expected that the Project Manager will manage specific internal communications processes.

- An effective communications strategy will plan the communication of the scope, objectives and progress of the project to the organization in order to achieve support for its implementation.
- Project success is reliant on effective communication to all stakeholders so that they understand their own responsibilities and those of others.
- A communication strategy is usually based on an understanding of how communication occurs within the organization on a specific topic. Often this is defined by a communications map (Figure 2-5) which can show both formal and informal lines of communication both within and external to a project.
- The overall strategy needs to be based on a stakeholder assessment (page 52) so that communication is a support mechanism for the project in harnessing stakeholder support and engagement.

Short case study

A new product is being introduced into a stable marketplace by the company with the largest share in the market. It is critical that the product launch is successful and delivers further market share. There are so

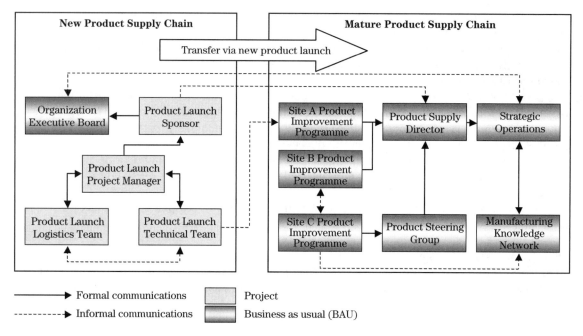

Figure 2-5 Example communications map

many distinct parts of the organization involved in the product supply chain that a communications map (Figure 2-5) was developed as a basis for the product launch communications plan.

Figure 2-5 shows the communications routes between the main business units from the perspective of the project. The lines shown represent formal communication with the arrows indicating the usual direction of that communication.

The map highlighted three areas which needed further investigation in order to fully understand the decision-making authority within each group, as applicable to the product launch project.

Strategic operations: This remains responsible and accountable for the manufacture and supply of product for all global markets. Within this business unit there are standard processes/people in place to effectively manage this part of the business:

➤ *Product portfolio management* – The organization has a supply director for each main product type. This person is accountable for the capability of the manufacturing site to meet the market demand and therefore responsible for managing capacity between the three sites which have the capability to make the existing and new product.

➤ *Site sales and operational planning processes* – There is an organizational process responsible for the internal site capacity prioritization whilst not impacting external capability.

Product steering group: The role is unclear and disconnected from main parts of each supply chain. It needs to be redefined or removed.

Manufacturing knowledge network: This will focus on strategic and tactical projects to provide the capability to manufacture all products to the highest standards of quality and efficiency. This means that there will need to be new communications routes between it and all site improvement activities.

As a result of development and discussion of the communications map the roles of key external stakeholder groups were better aligned and clarified and an effective communications strategy developed. This enabled a more supportive and informed external environment for the launch project and ultimately supported a successful new product launch and transfer into the mature supply chain.

Communications planning

Communications planning is therefore a part of the project business plan and should be completed so that internal and external communications are effective and aligned.

The key aims of a project communication plan are:

➤ To identify the audiences and ensure that their communication needs are met.
➤ To build broad awareness about the project among the stakeholder groups across all businesses and functions within an organization.
➤ To communicate relevant information to all those impacted by the project.
➤ To motivate stakeholders to devolve communication to all levels in the business (so those impacted understand the rationale for the project and can access quality information enabling them to be motivated by the project themselves).
➤ To promote good communications skills and processes within an organization as integral to the delivery of projects.

The delivery objectives are:

➤ To develop communications messages, materials and vehicles which provide information and support, plus news on a regular basis.

- To maximize the use of existing communications vehicles where possible (e.g. intranet, newsletters, town hall meetings).
- To recognize that the communication needs related to any project may vary due to different stakeholders, different project phases and current project progress.

Communications planning is a key part of the management of the link to the business and to the effective management of business change (page 47).

Communications need to be planned to:

- Get the audiences ready for the business change which the project will deliver. This is a process called 'unfreezing' which is a part of managing effective business change.
- Keep the audiences engaged during the business change and therefore during the project. This is about the management of transition from the previous to the new business environment (page 54).
- Get the audiences ready for the completion of the business change which is linked to the completion of the project. This is a process called 'refreezing', the final stage in business change management.

The input–process–output (IPO) diagram shown in Figure 2-6 demonstrates the process which would typically be followed.

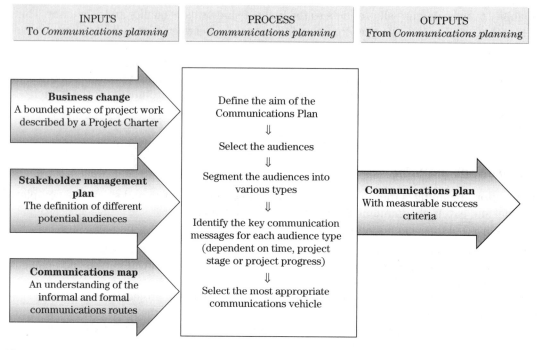

Figure 2-6 The communications planning IPO

Tool: Communications Planning Tool

There are many ways that the information can be collated once the communications strategy has been developed and this tool (Table 2-8) is very typical in that it relies on:

➤ Segmentation of stakeholders into categories so that each category can be considered separately in terms of their communications needs.
➤ Definition of communication objectives as linked to specific audiences or specific phases in a project.

This tool would therefore be developed in parallel with the stakeholder management plan (page 52) so that communications activities are aligned with other stakeholder management activities. Appropriate communication is an enabler of project success and should be well planned to coincide with key project milestones. There should also be an element of communication contingency planning so that appropriate communications occur when unexpected positive or negative project outcomes occur.

Table 2-8 Communications Planning Tool explained

Planning Toolkit – Communications Planning Tool			
Project: <insert project title>		**Project Manager:** <insert name>	
Date: <insert date>		**Page:** 1 of 1	
Measures of success			
<insert SMART measures which this communications plan is aiming to achieve – usually related to achievement of stakeholder management objectives and also linked to support of the project vision of success>			
Project phase	**Audience**	**Communications objective**	**Communications activity**
<insert the specific project phase from the project roadmap>	<insert the category of stakeholder or each individual stakeholder>	<insert the key message which is being communicated>	<insert the method by which the key message will be communicated>

Measures of success

These are quantifiable measures which demonstrate that the Communications Plan was the right plan delivered successfully. For example:

➤ Stakeholder feedback is positive: when asked 'do you understand the goals of this project and how it impacts you/your role within the organization?' more than 80% respond positively.
➤ 1 item of unsolicited positive feedback per month.
➤ 90% of stakeholders behave in a positive, aligning manner.

Project phase

Every project has a distinct roadmap which it follows from the start to the end (page 106). Most project roadmaps are made up of distinct phases which tend to require a different communications focus. For example:

➤ *Project launch* – At this stage it is usually critical that all stakeholders gain a broad, consistent understanding of the project and what it aims to achieve.
➤ *Design 'freeze'* – Some stakeholders may need a detailed understanding of the completed design whilst having no say in the actual decision; others may only need information on the completion of the stage.

Each project phase may also have a specific risk profile which determines the type of communications occurring in specific circumstances.

Audience

A focused, customized communications plan should develop specific messages for a specified audience. Typically the audience is categorized through stakeholder analysis (page 52).

Communications objective

Broad, unfocussed communications can do the very opposite of the intent. Each communications opportunity or event should be based around a clear message or set of messages. The clarity of communications objectives can then be tested when gaining feedback from the audience (solicited or unsolicited). Examples of communication objectives are:

➤ To generate a shared understanding of the project.
➤ To prepare people for the change to be implemented (the output of the project).
➤ To develop a consistent awareness of the current status of the project.

Communications activity

The final decision to be taken is the one that most plans get wrong: deciding how to deliver the chosen message to the selected audience (Table 2-9).

Examples of communications plans are included in the case study in this chapter (page 63), and in Chapters 6–8.

Table 2-9 Communications vehicles

Communication vehicle	Reasons for selection	Reasons for non-selection
1-1 meeting	- Personal delivery with immediate feedback on message effectiveness - Potential to adapt the message to suit individual needs – useful when the message has an impact on an individual	- Potential for inconsistency of message even if the same person delivers the message - It is very time inefficient if the audience category is large
'Town Hall' meeting	- Delivery of a consistent broad message to a large group of people - Able to get across the leadership position and show support	- Potential mix of different audiences and therefore different needs - Lengthy preparation time to maximize effectiveness - Mainly useful for broad messages
Focus group session	- Delivery of a consistent message to a small, select group of people - Two-way process with immediate feedback and generation of project ideas - Ability to ensure that similar stakeholders involved	- Level of interaction with the audience can make this a difficult vehicle unless very prepared - Not ideal for large stakeholder groups unless there is a rationale for sampling and follow on communications
Notice board (physical or electronic)	- Delivery of a consistent message to a selected group of people - Easily accessible as a part of BAU	- Not useful for 'big' messages or ones with great individual impact - Need an area where selected audience go as a part of BAU
Email	- Personal delivery to an individual's PC - Ability to check if message has been received/read	- Not useful for 'big' messages or ones with great individual impact - Not suitable within a high-volume email culture
Newsletter	- Delivery of a consistent message to a large group of people - Ability to put the message into an accessible format and style	- Lengthy preparation time to maximize effectiveness - Mainly useful for broad, high-level messages to a selected audience(s)

Benefits management

In order to be sure that the project is going to be a success there needs to be a clear link between the things the project will deliver and the benefits this will enable.

- The project business case defines the expectations the business has: 'why' the business needs the project to be delivered.
- The PDP details:
  - 'What' exact benefits will be delivered (the specification of the benefits).
  - 'When' and 'how' benefits will be delivered (benefits realization planning).
  - 'What' needs to change within the business in order to support the benefits and how sustainability is measured (business change and sustainability planning).

There is clearly a strong link between benefits management and business change management. Inevitably a project will have an impact on the business into which it will fit and may even need some supporting activities to enable benefits realization.

The benefits specification and realization planning will inform the business change and sustainability planning (page 47).

The section of the benefits management lifecycle (Figure 2-7) relevant to project delivery planning is the middle step – benefits management.

This is where we:

- Specify the benefits
  - Linking project scope to benefit criteria.
  - Linking benefit metrics to benefit criteria.
  - Set benefit metric targets based on business need.
- Define how to realize the benefits
  - Review the baseline for each benefit metric.
  - Define the size of the 'benefits gap' which the project, in partnership with the business, needs to fill.
  - Linking the benefits realization plan to the business change plan (page 49).
  - Linking the benefits realization plan to the sustainability plan (page 55).

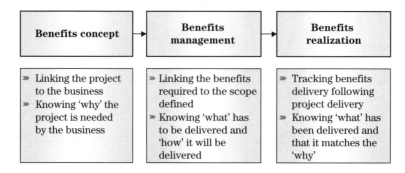

Figure 2-7 The benefits management lifecycle

Business case

In terms of benefits management, the business case represents the first stage in the benefits management lifecycle (Figure 2-7) as it defines the broad benefits concept which the project needs to align to. The benefits concept can usually be summarized through use of a simple benefits hierarchy (Appendix 9-4) which shows how the project definition links with both the project cost/benefits analysis as well as the organizational strategy. This analysis ensures that there is a robust rationale for the project: 'why' it is needed by the business.

Short case study

A product is nearing maturity in the marketplace and needs to be further developed if the company wants to maintain its current market share. A project to develop an improved product is proposed based on a clear benefits concept (Figure 2-8) which is a part of the approved business case:

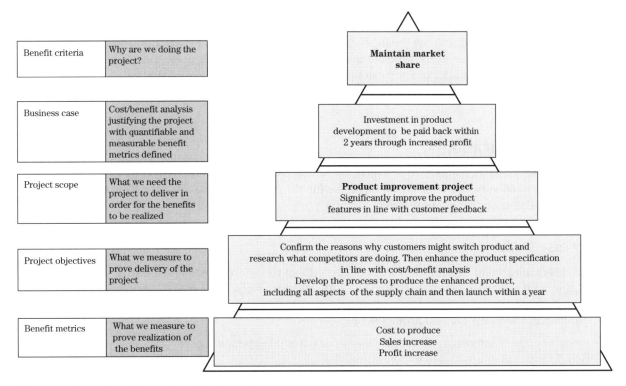

Figure 2-8 Example of Simple Benefits Hierarchy

- Develop the product specification so that customers are more likely to remain with this product than change to another.
- Ensure that any invested monies can be paid back within 2 years through increased profit margins.

The Simple Benefits Hierarchy was the basis for benefits specification work within the project and benefits metrics were eventually specified as:

- *Optimized cost of production* – minimize the production costs for the improvements and where possible decrease production costs for mature parts of the supply chain.
- *Increased sales revenue* – through effective product research aim to maintain current customers and to persuade those currently choosing a competitor product to switch. Through this metric market share would be tracked.
- *Increased profit* – profit is the difference between selling revenue and production costs. If revenue increases and production cost decreases then profit will increase (clearly it will only need a positive trend in one or the other to do this).

Business case issues

When a Project Manager begins to develop the PDP it is likely that one of the first documents he will receive is the business case. However this does not necessarily fully define the scope or the methodology of delivery, merely the business benefits which the organization requires for a specified level of investment. In addition the business case document for different types of project can have differing levels of detail due to the types of investments needed (capital or revenue, assets or people) as well as different organizational approval processes. Whatever the status of this key document it is the responsibility of the Project Manager to completely understand the business case he is contracted to deliver.

Benefits specification

During the early definition stages of a project there will be work completed which links specific areas of project scope to specific benefits, based on achieving the agreed benefits concept. This work can then be used to define specific benefit metrics.

Short case study

A facilities management organization was concerned that its external expenditure was 'out of control' and through investigation ascertained that the root cause was a lack of consistent operation when buying external supplies and services. A project was therefore approved based on this very general benefit concept and very general project scope:

- Project scope = improve internal processes for external expenditure.
- Benefit concept = controlled expenditure.

The resulting project benefits and scope mapping session supported both the benefits definition as well as scope definition (Figure 2-9) and enabled a clearer picture of priority in terms of both project objectives and benefits realization.

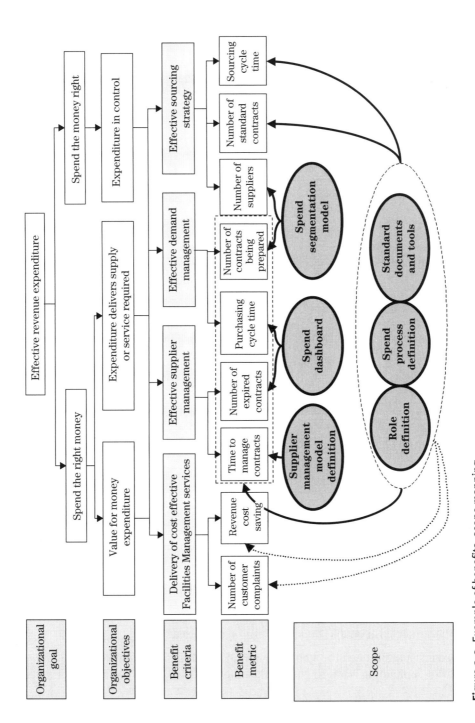

Figure 2-9 Example of benefits–scope mapping

The Benefits Specification Table Tool (Appendix 9-5) was then used to formalize this work for use in realization planning (Table 2-10).

Table 2-10 Example Benefits Specification Table

Project Management Toolkit – Benefits Specification Table						
Project:	*<insert project title>*		**Date:**	*<insert date>*		
Potential benefit	**Benefit metric**	**Benefit metric baseline**	**Accountability**	**Benefit metric target**	**Area of activity**	
What the project will enable the business to deliver	*Characteristic to be measured*	*Current level of performance*	*Person accountable for delivery of the benefit to target*	*Required performance to achieve overall benefits*	*The project scope that will enable this benefit to be delivered*	
Controlled expenditure	Time to manage contracts	To be completed during benefits realization planning activity		Reduce by 10%	Process, roles, standards and tools	Supplier model
	Number of expired contracts			Eliminate		Spend dashboard
	Purchasing cycle time			Reduce to 3 months		
	Number of suppliers			Reduce by 25%		Spend segmentation
	Cost savings			Dependent on cycle time		Launching new process ASAP

Benefits realization planning

Once the benefits have been fully specified there needs to be a plan for their delivery. This plan should answer all the following questions:

➤ Are there baseline and target measures for all identified benefits metrics?
➤ Who is responsible for each individual metric? How will they track realization of the benefit to the agreed target level?
➤ What part of the project will enable the benefit to be delivered?

Within a PDP there is usually a need for a collation of all benefits definition work completed to date, as well as some additional planning work to ensure that the benefits can be realistically realized.

Tool: Benefits Realization Plan

Although the Benefits Specification Table Tool (Appendix 9-5) can be used to address most aspects of a benefits realization plan, there is value in using Table 2-11 as a collation tool. This also ensures continuity between the various phases of benefits definition and planning.

Table 2-11 Benefits Realization Plan Tool explained

Planning Toolkit – Realization Plan				
Project: <insert project title>		**Project Manager:** <insert name>		
Date: <insert date>		**Page:** 1 of 1		
Benefits concept				
<insert a copy of the Simple Benefits Hierarchy>				
Benefits specification				
<insert a copy of the Benefits Specification Table>				
Benefits realization				
Benefit metric	**Tracking frequency**	**Dependency**		**Priority vs vision of success**
		Project scope	**Business change**	
<insert benefit metric to be tracked>	<insert when tracking starts and at what frequency>	<insert any scope items, without which the benefit metric will not improve>	<insert any change external to the project, without which the benefit metric will not improve>	<insert criticality of this benefit metric to the overall achievement of the vision of success>

Benefits concept

The plan should contain a reference to the approved benefits concept. Bear in mind that this might be a detailed benefit statement linked to an investment value, or organizational objectives linked to the overall goal of an organization.

Benefits specification

If a full Benefits Specification Table is available it should be referenced here. Alternatively, at the start of the planning phases there may only be a list of required benefit metrics and their specified target levels.

Benefits realization

Actually planning to realize a set of specified benefits begins when there starts to be a causal link between critical areas of project scope and individual benefits. Benefits may also arise from a cumulative effect of a number of scope areas; however, they do not arise from doing nothing! For each benefit metric the following planning activities should therefore be completed:

- *Benefits metric*: The exact metric should be inserted including its units. Sometimes a benefit metric will be a combination of a number of different measures and the calculation should be made clear.

  For example if a utilization metric is to be used there need to be a clear definition: percentage utilization of an office based on people per floor area.

- *Tracking frequency*: Some benefits may start to be realized during the delivery of a project whereas others require the project to be fully completed before any change in benefit is likely. In this part of the tool there should be some indication of the anticipated start of benefits delivery and the overall duration to achieve the target value. Additionally the method of tracking and the frequency should be noted.

  For example a project to launch a new product will not be delivering the benefit metrics related to external sales revenue increase until after the product has been launched. On the other hand, a project to improve production line efficiency may start to see improvements during the implementation of the project.

- *Dependency – project scope*: The critical scope areas, without which the benefits cannot be realized, should be stated including any specific level (scope quality, quantity and/or functionality).

- *Dependency – business change*: A benefit may require both a specific area of project scope to be completed and some additional change within the business which is outside of the scope of the project.

  For example, an improvement in the efficiency of a new laboratory may be reliant on completion of a new laboratory layout and operating procedures (project scope), and laboratory analyst recruitment and training (excluded from project scope). Business changes need to be clearly articulated so that all stakeholders are clear on what needs to be achieved within and external to a project.

- *Priority vs vision of success* – Through the development of the scope and associated understanding of its link to benefits delivery, there are often secondary benefits realized by virtue of a combination of scope and/or a combination of benefits. What is important is that the highest priority benefits are identified in terms of the definition of project success.

  For example, in Table 2-10 a number of benefit metrics were identified:
  - *Time to manage contracts* – a high priority for the sponsor. Currently the root cause of the lack of control is seen in ineffective management of contracts (through trying to manage too many contracts and contract issues at once).
  - *Number of expired contracts* – a high priority for the sponsor. When a contract has expired the company is at risk and this lack of control is not acceptable.
  - *Purchasing cycle time* – a high priority for the sponsor. Every additional day it takes to complete the contract costs the organization (lost savings, additional management time, more expires, delivery inefficiencies for the customers and therefore potential complaints).
  - *Number of suppliers* – a medium priority for the sponsor. Although supplier rationalization is another organizational goal it is not the focus of this project. However it is highly likely that the new processes will have a positive impact.
  - *Cost savings* – a low priority for the sponsor. Although cost reduction is an organizational goal it is not the focus of this project. However it is likely that the new processes will have a positive impact.

Business change management

Every project will impact a business in some way. It will cause or be a part of a business change (Figure 2-10). During the development of the PDP it is appropriate to form a fuller understanding of the needs of the project in regard to the business environment, and any changes that need to be made within that environment.

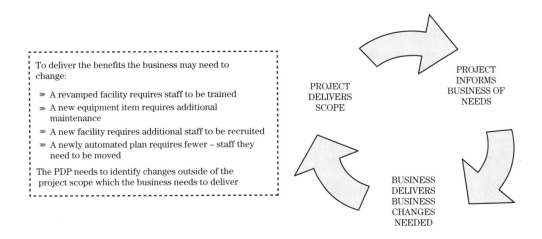

To deliver the benefits the business may need to change:

- A revamped facility requires staff to be trained
- A new equipment item requires additional maintenance
- A new facility requires additional staff to be recruited
- A newly automated plan requires fewer – staff they need to be moved

The PDP needs to identify changes outside of the project scope which the business needs to deliver

PROJECT
DELIVERS
SCOPE

PROJECT
INFORMS
BUSINESS OF
NEEDS

BUSINESS
DELIVERS
BUSINESS
CHANGES
NEEDED

Figure 2-10 Projects and business change

The business change process

Within most organizations business changes occur on a regular basis – some large step changes and some small incremental continuous improvements. All will go through the same process, often described as the change cycle (Figure 2-11).

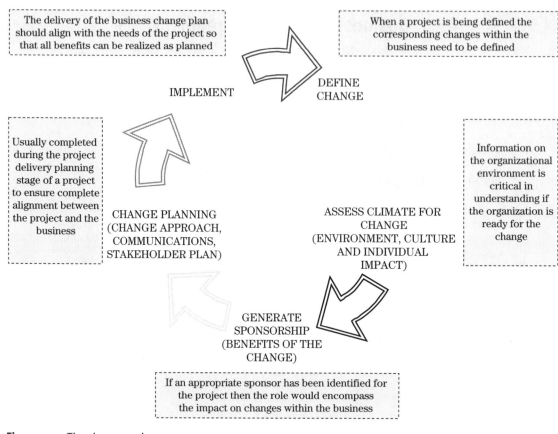

Figure 2-11 The change cycle

Figure 2-11 also indicates the interaction between the project and the business during the change cycle:

➤ When defining a project, the business environment (and any change within it) should be defined.
➤ A part of the development of the PDP is the development of the change plan and the integration of environmental assessment within it.
➤ A part of the generation of sponsorship for the change is creating stakeholder engagement and the project role within that change.

The management of the integration of the project within the business, and the sustainability of the benefits realized, is an additional necessary activity. This is called sustainability planning, a key part of the business change plan.

The business change plan

Typically an organization would formulate a business change plan (Figure 2-12) which would:

- Define all the changes to be made to the business. This could be through one or more projects and also one or more incremental or step changes.
- Understand the forces which may resist the change and therefore the risks of implementing the change.
- Identify all the organizational resources needed for the change to be successfully and sustainably delivered and how those resources are to be deployed.

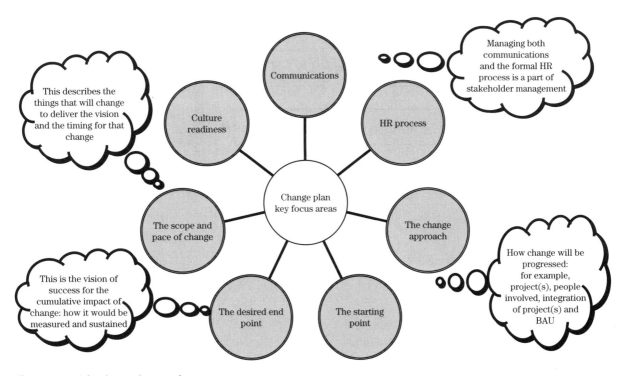

Figure 2-12 A business change plan

From the perspective of the project and the PDP, the change plan should be a support mechanism, but it may not always be in place if the project is the major part of the change. In this case the PDP should identify those elements of a typical change plan which need to be addressed by the business in order for the project to be successful.

The Benefits Realization Plan (page 45) will be fundamental to this process as it indicates which benefits require changes which are outside of the scope of the project. The following should also be a part of the PDP:

- Environmental assessment.
- Stakeholder analysis and planning.
- Sustainability planning.

Environmental assessment

In the majority of instances business changes will impact the current culture within an organization. Culture is developed as described by Figure 2-13:

➤ Culture = how we behave + values we hold + things we do.
➤ To change the culture we need to change these.

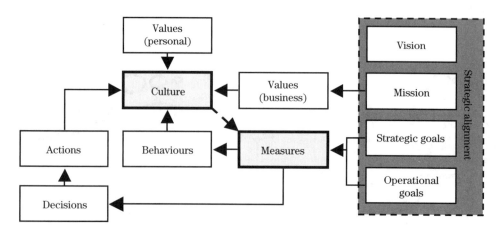

Figure 2-13 An organizational culture model

Prior to an organization starting a major change it is sensible to review the internal environment. This means assessing:

➤ Culture – the way we do things round here:
  ⊳ Formal – business processes, measures.
  ⊳ Informal – communication, values, behaviours.
➤ Cultural history – the ways things have been done in the past.
➤ Organizational direction – where we are aiming to go:
  ⊳ Business strategy/vision.
  ⊳ Strategic alignment.
  ⊳ Need for change.

This environmental assessment will give a view on readiness for change and therefore the likely success and/or appropriate strategies for success.

A simple model (Figure 2-14) can be used to assess readiness based on two environmental factors:

➤ Internal factor – history of change.
➤ External factor – drivers for change.

		PUSHED INTO CHANGE	READY FOR CHANGE
	HIGH	The organization has a poor history for change but needs to respond to key challenges in the environment. It tends to be static until the situation is high risk	The environmental analysis has demonstrated a 'burning bridge' and the organization accepts this and plans an appropriate change
Drivers for change	**LOW**	NO CHANGE REQUIRED This organization is static and can remain so in its current environment	READY FOR CONTINUOUS IMPROVEMENT This organization is always looking at opportunities to continuously improve and can successfully implement incremental change which sustains
		LOW SUCCESS	HIGH SUCCESS
		History of change	

Figure 2-14 Change readiness matrix

This model works because the strongest internal force is an organization's previous experience of change. This is reinforced by the need for change, usually an internal reaction to an external driver.

A Project Manager needs to understand whether there is resistance to the changes that the project will be making to a part of the business. It is useful to work with the sponsor and the core Project Team to brainstorm both factors and come to an agreement on which quadrant the business area is in.

For larger changes there is additional value in conducting a more detailed review against a broader range of environmental factors (Table 2-12).

Table 2-12 Typical environmental factors impacting change success

Environmental factor	Ready for change
➤ History of change	Many examples of successful organizational change
➤ Organizational strategy	Aligned strategy linked to a shared vision
➤ Business processes	Appropriate ways of working backed up with appropriate measures and levels of empowerment
➤ Communication	Open communication channels – horizontally and vertically
➤ Drivers for change	Clear 'burning bridge' for all levels in the organization

Other factors which could be considered are:

- Resilience – the ability of the organization and individuals to handle (further) change.
- Reward – the benefits from the change at an organizational, team and individual level.
- Respect – the potential for change to impact the self-respect of an individual.
- Disruption – the level of disruption to the status quo.

Customized questionnaires can be generated which assess the above factors and a profile of the organizational environment can be obtained. This supports the development of an appropriate change plan as it can highlight particular areas of resistance and/or support. It can also highlight which stakeholder groups require detailed analysis and then management.

Stakeholder analysis and planning

One aim of the business plan section within a PDP is to ensure that all relevant stakeholders are engaged and supportive of the project and are willing to play their part in its success. In order to do this the following process should be used:

- *Identify all stakeholders*: Whether the project is the main change or one change in a programme, it will impact groups of people, and groups of people will have the ability to impact its outcome. All various types need to be identified initially:
  - Review organizational charts.
  - Review the business areas which are impacted by the change.
  - Discuss the need for any additional stakeholders within the organization.
  - Consider if there are any stakeholders external to the organization.
- *Analyse stakeholders*: The level of support or resistance from stakeholders can impact the selected project delivery strategies. A simple stakeholder categorization model can be used (Figure 2-15):
  - Analyse the level of support from key stakeholders (who have power and/or influence).
  - Identify what could cause a stakeholder to move into a different category (either a positive or negative move).
  - Highlight which stakeholders are supportive, resistant or neutral to the change which the project will be making and/or to the project itself.
- *Develop a stakeholder management plan*: A key part of the role of the Project Manager (with support from the sponsor) is to ensure that the project has the appropriate level of stakeholder engagement for success (Appendix 9-6). This means:
  - Analysing how much the project needs the support of any particular stakeholder or particular type of stakeholder.
  - Identifying a strategy to achieve the required level of stakeholder engagement.
  - Identifying how each stakeholder is to be managed, who will take responsibility for that management and how success will be measured.

Using stakeholder mapping (Figure 2-15) can make the management plan more effective as it prioritizes those stakeholders whose support is critical to success.

For example, a resistant 'partner' is a bigger issue than a resistant 'acquaintance' and should be dealt with as a high priority. A strong stakeholder management plan is needed based on a clear understanding of what is driving the resistance and what would overcome it. A force-field analysis is a useful tool to use to understand this issue and can be helpful in identifying appropriate management strategies. Figure 2-16 shows an example for a force-field analysis of a senior stakeholder within a major organizational change programme.

		ASSOCIATES	PARTNERS
Importance (power/authority)	HIGH	We need to have appropriate contracts in place with our **associates** – they control resources and approve direction and pace	We need to build strong relationships with our **partners** – they can impact project progress and success, influence project direction and other stakeholders
		ACQUAINTANCES	FRIENDS
	LOW	We may communicate with our **acquaintances** on an 'as needs basis' and keep a watching brief on where they move within the matrix as the project progresses	We need to maintain an open relationship with our **friends** – they can influence what other more powerful stakeholders decide
		LOW	HIGH
		Influence	

Figure 2-15 Stakeholder mapping

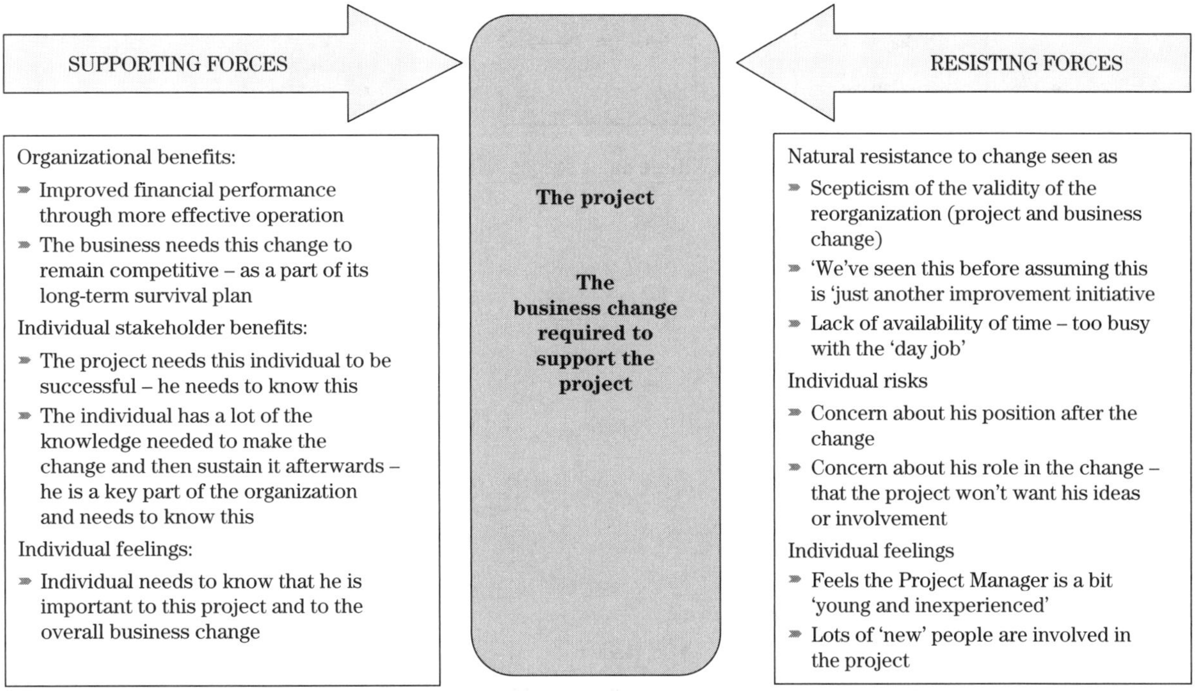

SUPPORTING FORCES ➤ | The project / The business change required to support the project | ⬅ RESISTING FORCES

Organizational benefits:
➤ Improved financial performance through more effective operation
➤ The business needs this change to remain competitive – as a part of its long-term survival plan

Individual stakeholder benefits:
➤ The project needs this individual to be successful – he needs to know this
➤ The individual has a lot of the knowledge needed to make the change and then sustain it afterwards – he is a key part of the organization and needs to know this

Individual feelings:
➤ Individual needs to know that he is important to this project and to the overall business change

Natural resistance to change seen as
➤ Scepticism of the validity of the reorganization (project and business change)
➤ 'We've seen this before assuming this is 'just another improvement initiative
➤ Lack of availability of time – too busy with the 'day job'

Individual risks
➤ Concern about his position after the change
➤ Concern about his role in the change – that the project won't want his ideas or involvement

Individual feelings
➤ Feels the Project Manager is a bit 'young and inexperienced'
➤ Lots of 'new' people are involved in the project

Figure 2-16 Stakeholder force-field analysis

As a result of this analysis the particular stakeholder was managed differently – as an individual rather than as a part of a group of similar stakeholders. He received individual updates and was kept informed regularly by the sponsor who was of equivalent standing in the organization. As a result the project received the support necessary and the stakeholder also 'got through the change' with fewer issues (for him, the project and the business).

A Project Manager should always remember that although he is dealing with major changes in a business, it is likely that people will see the changes from a very individual perspective. A stakeholder management plan (Appendix 9-6) needs to consider the three basic stages that any stakeholder will go through (Figure 2-17) and in particular how that stakeholder feels as the change is occurring (Figure 2-18).

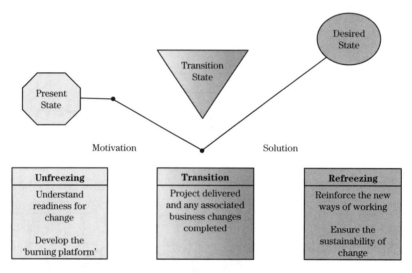

Figure 2-17 Change process – three basic stages

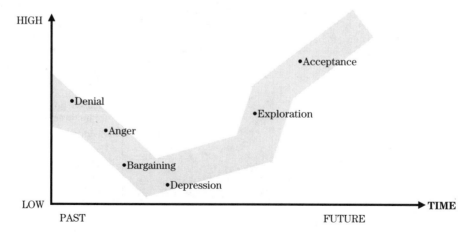

Figure 2-18 The change curve

The change curve (Figure 2-18) is useful in understanding the reactions of all stakeholders as the project progresses (the transition phase shown in Figure 2-17). Some stakeholders' roles post-project may be significantly impacted and so their response to the change may be more extreme than other stakeholders. Some may even be members of the Project Team and so their potential 'depression' needs to be managed. The stakeholder management plan needs to anticipate this.

Stakeholder management activities are usually focused around effective and appropriate communications. The project Communications Plan (page 34) must therefore be developed in line with the result of the stakeholder analysis and the defined stakeholder management plan.

The stakeholder management plan will also be a useful input to the development of the sustainability plan which aims to maintain the project vision of success once the change has been integrated into BAU.

Sustainability planning

The final part of business planning within a PDP is the development of a sustainability plan. A sustainability plan aims to ensure that the project, once complete, is handed over in such a way that benefits can be realized and all changes remain in place. Figure 2-19 summarizes the three main sections.

The reason that sustainability planning starts during PDP development is to ensure that the interface between the project and the business is clear.

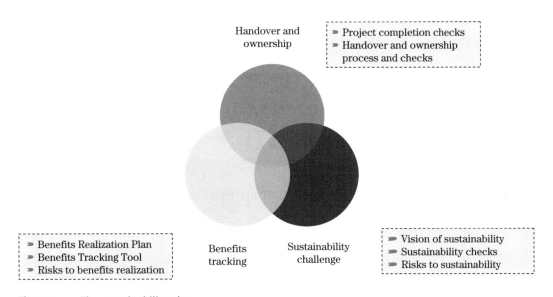

Figure 2-19 The sustainability plan

Handover and ownership

There should be an agreed project status in order for the customer to accept that it is complete and therefore take ownership.

Often this changes towards the end of the project as the customer accepts some unfinished deliverables; however at the PDP development stage, it is usual to state which project deliverables need to be at 100%.

This part of sustainability planning also allows the customer, sponsor and Project Manager to discuss the definition of ownership particularly with respect to project deliverables, which are crucial to sustainability, and benefits tracking.

Benefits tracking

The Benefits Realization Plan should be used to define the level of benefits tracking which is to be completed after project handover.

At this stage the customer may already be considering which benefit metrics will be incorporated into BAU as operational metrics, to be used on a continuous basis to run the business.

There should also be a risk assessment against each benefit metric target:

➡ What will stop the benefit reaching its target value?
➡ How can high risks be mitigated?
➡ What is the contingency plan should the risk occur so that the benefit target can still be achieved?

Sustainability challenge

The sustainability challenge should be used to define some explicit checks versus the agreed vision of a sustainable change. A sustainability check needs to remain within a target range in order not to prevent the realization of benefit(s). Sustainability checks are not usually incorporated into BAU as operational metrics as they should only be checking whether changes (post-project completion) have been maintained. There should also be a risk assessment against the vision of a sustainable change:

➡ What will stop the sustainability checks being positive?
➡ What will stop the vision of sustainable change being achieved?
➡ How can high risks be mitigated?
➡ What is the contingency plan should the risk occur so that the sustainability check can still be positive?

Tool: The Sustainability Plan

Typically all sustainability planning issues would be brought together into one document: a Sustainability Plan. Table 2-13 is an example of this.

Table 2-13 Sustainability Plan Tool explained

Planning Toolkit – Sustainability Plan			
Project: *<insert project title>*		**Project Manager:** *<insert name>*	
Date: *<insert date>*		**Page:** 1 of 1	
Vision of success			
<insert vision of sustainability so that it can be tangibly measured>			
Scope handover and ownership			
Final deliverable	**Status**	**Handover process**	**Ownership**
<insert project deliverable>	*<insert required status at handover>*	*<insert process by which deliverable will be owned by the customer>*	*<insert owner – name and role>*
Benefits realization			
<insert Benefits Tracking Tool>		*<insert benefits risk assessment>*	
Sustainability			
<insert sustainability checklist>		*<insert sustainability risk assessment>*	

Vision of success

As a result of the project there should be some stated vision which describes the sustainable output from the integration of the project and other associated business changes.

For example, a project to upgrade a manufacturing facility to the most current technology may have the following vision of sustainable change – '*this manufacturing facility will operate state of the art technology to deliver the required capacity and quality of product at a best in class cost, delivering to the customer on time every time*'.

The vision should be one that is possible and plausible when considering what the project will deliver (an upgraded facility with increased capacity and process cycle time) and what the business will be changing upon receipt of the final project deliverable (reduced manpower costs, reduced process waste, increased energy costs, trained operators, new standard operating procedures (SOPs)).

The vision should be worded so that its achievement can be measured. This supports definition of appropriate sustainability checks and confirms the benefits anticipated from the completed change.

Scope handover and ownership

This section should be used to list all the final deliverables and confirm specific handover criteria and ownership issues for each item.:

➤ *Final deliverable*: List the tangible items to be handed over, such as assets and documents.
➤ *Status*: For each deliverable describe the status for handover, for example an engineering drawing may need to be 'as built' and a piece of equipment may need to be fully tested to a specified performance level.
➤ *Handover process*: Describe how each deliverable should be handed over, for example a drawing may need to be signed-off as acceptable and a new asset may need to be physically inspected.
➤ *Ownership*: Not all deliverables will be owned by the same person in an organization although there will be one overall customer. For example the engineering department may take ownership of the assets in terms of maintenance but operational ownership is with the plant manager.

Benefits realization

This section should be used to confirm the benefits management plan.

➤ *Benefits Tracking Tool*: Attach the proposed Benefits Tracking Tool (Appendix 9-7), showing what needs to be measured and when.
➤ *Benefits risk assessment*: Use a Risk Assessment Tool (Chapter 4, page 157) to identify and analyse the risks to achieving each benefits metric. Identify a mitigation action plan.

Sustainability

This section should be used to confirm the method of maintaining sustainability of the changes.

➤ *Sustainability Checklist* – This tool identifies checks which enable achievement of the benefit metric targets (Appendix 9-8). A causal relationship between a change that has been delivered and a benefit which is to be realized needs to be made. The check then identifies if the change has been sustained and therefore whether the benefit will be sustainably realized.
➤ *Sustainability risk assessment*: Use a Risk Assessment Tool (Chapter 4, page 157) to identify and analyse the risks to achieving each Sustainability check. Identify a mitigation action plan.

Examples of sustainability plans are included in the case studies in Chapters 6–8.

Business plan case study – production capacity improvement

To illustrate the key points from this chapter an extract from the PDP for a manufacturing improvement project follows.

Situation

A pharmaceutical device manufacturing facility is facing a crisis:

- Reject rates are increasing.
- Work in progress (WIP) inventory appears completely out of control.
- There are frequent production line stoppages due to a lack of available material to run the line.
- An older production line appears to be operating at better yields than the two newer ones which have never achieved stated design capacity.
- Production lead times are growing and customers are receiving orders later and later.

In addition current sales are forecast to double in the next 6 months as the device is launched into new markets. The reaction from the business was to immediately commence a project to increase manufacturing capacity through the addition of more production equipment and resources. A business case was developed to support this (Table 2-14).

Table 2-14 Approved business case – capacity improvement project

Capacity Improvement Project Business Case			
Business case developed by:	Director of Device Manufacturing Strategy	**Date:**	January 1st
Project reference number	PR_059	**Business area**	Manufacturing strategy
Project Manager	To be confirmed	**Project sponsor**	Director of Device Manufacturing Strategy
Business background	Within 3–6 months the sales forecast for the device will double from 90 to 180 packs/month (each pack contains 500 devices) Current output matches demand through the use of one older production line (line 2) and two newer production lines (lines 1 and 3)		
Project description	The installation of an additional production line (design capacity of 100 packs/month)		
Delivery analysis	Use of site based project engineers to procure the production line (use preferred supplier) and install in line with all appropriate regulations		

(continued)

Table 2-14 (Continued)

Capacity Improvement Project Business Case			
Business case developed by:	Director of Device Manufacturing Strategy	**Date:**	January 1st
Project reference number	PR_059	**Business area**	Manufacturing strategy
Project Manager	To be confirmed	**Project sponsor**	Director of Device Manufacturing Strategy
Business change analysis	The project is required to support the launch of the device into new markets in line with the business strategy for this product line		
Value-add analysis	Cost of investment is $1million (25% accuracy) with a 4 month lead time for the production line delivery to site. Revenue investment is also required in terms of additional operations staff and other site infrastructure costs. Anticipated returns are additional sales of 90 packs/month worth an estimated $250,000		
Impact of NOT doing the project	If the additional capacity is not delivered then the new markets cannot be supplied		
Project approved (Value Add or Not?)	YES	**Name of approver and date**	Vice President Manufacturing

Business plan

Immediately following approval of the project an experienced Project Manager was appointed from within the site engineering team. She reviewed the business case and established that the project goal was to 'increase additional production capacity to meet market needs'. On this basis she began the process of project delivery planning commencing with the business plan.

Sponsorship

The Project Manager quickly established a working relationship with the sponsor (Table 2-15), the Director of Manufacturing Strategy and developed a working stakeholder map in the form of a power–influence matrix (Figure 2-20).

Table 2-15 Sponsor Contracting Checklist – capacity improvement project

Planning Toolkit – Stakeholder Contracting Checklist			
Project:	Capacity Improvement Project	**Project Manager:**	Jane Jones
Date:	February 1st	**Page:**	1 of 1

Contract set-up

Stakeholder name – Director of Manufacturing Strategy, Peter Smith
Link to this project – Peter is the assigned sponsor and the appropriate one considering his role and the potential changes within the business which could result from this project

Apart from significant authority over this project, Peter is very well connected within the business. He has worked with the site many times before and was previously the Site Manufacturing Director. He supported the appointment of the current person in that role and is well respected by all on-site

Stakeholder background

Likely stakeholder engagement on this project – Peter is very motivated to deliver the benefits concept and is open to new ideas
Likely stakeholder knowledge of this project – Sponsor developed the approved business case based on his understanding that capacity could only be improved by additional, new production equipment

Project requirements from stakeholder

List any information the project will need from this stakeholder – Peter needs to share information on the other stakeholders as he is very familiar with the history of this facility
List any decisions the project will need from this stakeholder – Peter needs to approve the project charter and then support the Project Manager in selling it to other senior stakeholders. Sponsor will then approve the PDP prior to delivery (linked to final funding approval)
List likely impact this stakeholder may have on the project outcome – Peter has both power and influence and if he doesn't believe that the project will deliver he will block it. Therefore need to provide robust and compelling data to support any proposal

Stakeholder requirements from project

List any information this stakeholder will need from the project – Peter will want to be kept informed of the various options and likely potential for each to deliver the benefits. He will need regular updates
List likely impact this project may have on the stakeholder business operations – This project is at the core of Peter's business responsibilities
List any current stakeholder concerns – Peter wants a quick, efficient project so he will be concerned at any stalling – if new production equipment is not the right answer then he needs to know what is

Contract agreement

Outline what level of communication has been agreed (throughout the project) – Weekly face-to-face until PDP approved and then every 2 weeks until project completion
Confirm that the contract above (decision and information flow) has been agreed between Project Manager and stakeholder – Explicit contract agreed and no concerns regarding maintaining engagement

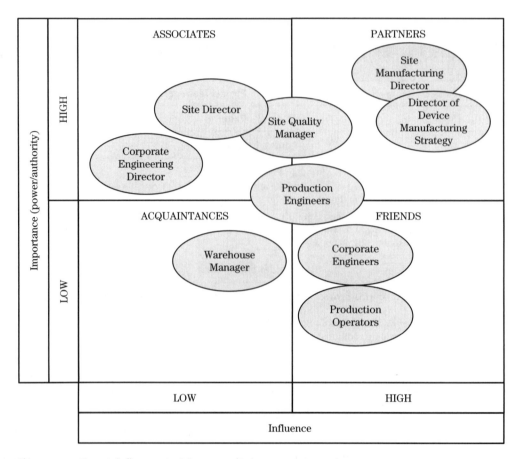

Figure 2-20 Power–influence matrix – capacity improvement project

This enabled the development of a customized communications plan. The communications plan (Table 2-16) was developed in the knowledge that the business case (Table 2-14) had generated a solution to an identified problem but that all viable options had yet to be considered. The first formal communication was the project charter (Table 2-17) and the goal of this communication was to open up stakeholder expectations to a different solution.

Table 2-16 Communications Plan – capacity improvement project

Planning Toolkit – Communications Plan			
Project: Capacity Improvement Project		**Project Manager:** Jane Jones	
Date: February 8th		**Page:** 1 of 1	
Measures of success			
The main aim of the Communications Plan is to engage stakeholder support so that the RIGHT project can be delivered: one which delivers the required benefits at the optimum organizational cost. KPIs being tracked are: ➤ Stakeholders with high power make timely decisions which support project progress ➤ Stakeholders with high influence are overtly supportive of project plans			
Project phase	**Audience**	**Communications objective**	**Communications activity**
Project launch	Partners	➤ Gain agreement on project aim and confirm benefits delivery ➤ Gain agreement on handover and business change issues – how it fits with BAU	➤ 1-1 meeting based around DRAFT project charter
Project delivery		➤ Confirm project status, it is on track and in control, risks are being managed, some benefits are already being delivered	➤ 1-1 meeting every 2 weeks ➤ Monthly one-page bulletin and risk review meeting
Project close-out		➤ Project has been delivered and is ready for handover for sustainable benefits delivery	➤ Close-out meeting based around sustainability plan
Project launch	Associates	➤ Gain agreement on project aim and confirm benefits delivery and any handover issues	➤ 1-1 meeting based around DRAFT project charter
Project delivery		➤ Project is on track to meet benefits delivery goal and handover needs	➤ 1-1 meeting every month using one-page bulletin
Project launch	Friends	➤ This is what the project is all about and how it integrates into BAU	➤ 1-1 meeting based around project charter
Project delivery		➤ Project is on track to meet benefits delivery goal	➤ Monthly one-page email bulletin and ad hoc meetings
All phases	Acquaintances	➤ Current project status – we are on track and not impacting your area	➤ Monthly one-page email bulletin

Table 2-17 Project Charter – capacity improvement project

Planning Toolkit – **Project Charter**			
Project:	Capacity Improvement Project	**Project Manager:**	Jane Jones
Date:	February 8th	**Page:**	1 of 2
Project description		**Project delivery**	

Project description

Sponsor
Director of Device Manufacturing Strategy

Customer
Site Manufacturing Director

Project aim
Increase production capacity to meet market needs

Project objectives
- Observe the current operation of the 3 production lines and collect performance data
- Conduct 3 kaizen workshop events to design and implement appropriate change on each line
- Modify equipment (as needed)
- Develop SOPs to support sustainability of changes (including performance measures)

Benefits
- Line capacity increased on each line so that total capacity meets forecast (180 packs/month)
- Minimal operational running cost increase
- Improved product quality
- Improved process cycle time and therefore ability to meet customer lead times

Final deliverable
- Three operational lines of a specified capacity
- New SOP and Performance Measures Board

Project delivery

Project Team
- There will need to be a core team and then three separate kaizen teams
- The kaizen teams will be made up of ALL of the line team with each line manager being in the core team
- Equipment engineers will also be available
- Warehouse manager

Additional resources
- An external consultant is needed to support the process improvement workshops (kaizen) using the lean six sigma methodology
- A revenue budget of approximately $100000
- Each line will need to be 'off-line' for the kaizen week and potentially for the following week to complete modifications

Critical success factors
- Engagement of the line managers and operators
- Accurate data on current operations
- Understanding of complete supply chain process: from customer order, to despatch, to customer
- Ability to modify operation of the line, physical layout of the line and the way that line teams work
- Management support and sponsorship
- Access to lean six sigma expertise

(continued)

Table 2-17 (Continued)

Planning Toolkit – Project Charter			
Project:	Capacity Improvement Project	**Project Manager:**	Jane Jones
Date:	February 8th	**Page:**	2 of 2
Project description		**Project delivery**	

Project description	**Project delivery**
Interim deliverables ➤ Process observation check sheets ➤ Kaizen workshop design ➤ Line teams trained in kaizen tools and methodology ➤ Kaizen output × 3 ➤ Equipment layout modifications ➤ Equipment modifications ➤ Pilot SOP and performance measures	**Risk profile** ➤ Root cause for poor line performance may not be easily solved ➤ Line teams may be defensive and not want to support the changes ➤ Other parts of the supply chain may not support the changes **Organizational dependencies** ➤ Customer orders – volume and timing ➤ Warehouse capacity
Critical milestones vs deliverables ➤ End January – all process data ➤ March 1st – line 1 kaizen (pilot) ➤ April 15th – line 2 kaizen ➤ May 5th – line 3 kaizen ➤ June 1st – handover to BAU	**Project delivery approach** The strategy for this project is to pilot an improvement process on one line as a pilot and then to use it on the 3 lines The process being followed assumes that the root cause problem can be solved by process changes and minor equipment modifications. This is on the basis that the lines are operating at less than 50% of their design capacity.

The Project Manager had to provide evidence that the project approach was highly likely to deliver the required benefits, which she did based on process improvements at a similar manufacturing plant in the UK.

Benefits management

In attempting to understand the issue to be resolved, the Project Manager facilitated a benefits mapping session with a group of senior stakeholders. The high-priority benefits identified were:

➤ *Ability to meet customer orders on time* – with implied capacity and supply chain speed requirements.
➤ *Product quality right first time* – thus reducing the cost of poor quality.
➤ *Profitability of the manufacturing operation* – so that the department remained a viable part of the long-term future of the organization.

As a result of this session stakeholders were open to three options for delivering the required set of benefits:

- *Option* 1: Leave current production lines as they are and buy new a new line.
- *Option* 2: Improve current production lines and buy a new line.
- *Option* 3: Improve current production lines.

The decision regarding which option to choose was then made against an agreed set of criteria (Table 2-18).

Table 2-18 Option decision matrix – capacity improvement project

Criteria	Option 1	Option 2	Option 3
Ability to meet capacity needs within 6 months	**Yes – high risk** Buy one large line	**Yes – high risk** Buy one small line	**Yes – medium risk** Improve all three lines to some extent (evidence to show this is possible)
Reliability of solution (sustainable capacity increase)	**Medium** New line may not meet design capacity, root cause problems not solved	**Medium** New line may not meet design capacity and older lines may be hard to improve sustainably	**Medium** Older line may be less reliable than newer ones but root cause problems would be targeted
Maintain cost of goods at current level (cost per capacity)	**No** Capital investment payback and increased costs of operating an additional line	**No** Investment payback and increased operating costs off-set by improvements	**Yes** No additional running costs and potential to reduce through improvements
Improve manufacturing yield/product quality	**No** Does not address root cause of current issues	**Yes** Potential to address some of issues	**Yes** Improved capacity will be achieved by reducing waste and releasing flow
DECISION	**NO**	**NO**	**YES**

- Option 1 was rejected as it was unable to meet the majority of benefits criteria:
  - Lines are likely to be fully operational too late to meet customer needs.
  - It is highly likely that quality and capacity problems would also occur with the new line.
- Option 2 was rejected as its divided focus was likely to be too much of a distraction for the team and the ongoing operations at the facility.
- Option 3 was selected as it met all high-priority benefits criteria even though it was recognized that there was still an element of risk in going with this option.

Based on this review of the ability to deliver the highest-priority benefits for the business, all stakeholders agreed that Option 3 should be progressed. The draft Project Charter was agreed and approved by all those with high authority and influence. The Benefits Realization Plan (Table 2-19) was also developed and approved.

Table 2-19 Benefits Realization Plan – Capacity Improvement Project

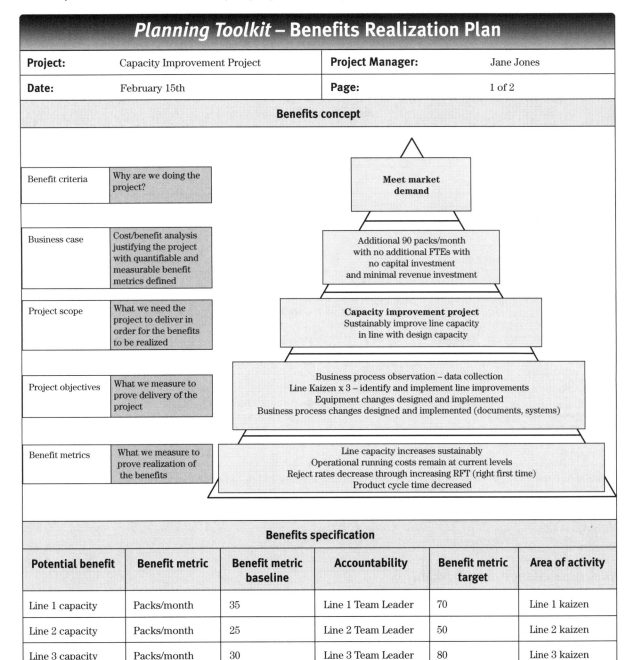

Planning Toolkit – Benefits Realization Plan

Project:	Capacity Improvement Project	Project Manager:	Jane Jones
Date:	February 15th	Page:	1 of 2

Benefits concept

Benefit criteria	Why are we doing the project?
Business case	Cost/benefit analysis justifying the project with quantifiable and measurable benefit metrics defined
Project scope	What we need the project to deliver in order for the benefits to be realized
Project objectives	What we measure to prove delivery of the project
Benefit metrics	What we measure to prove realization of the benefits

Meet market demand

Additional 90 packs/month with no additional FTEs with no capital investment and minimal revenue investment

Capacity improvement project
Sustainably improve line capacity in line with design capacity

Business process observation – data collection
Line Kaizen x 3 – identify and implement line improvements
Equipment changes designed and implemented
Business process changes designed and implemented (documents, systems)

Line capacity increases sustainably
Operational running costs remain at current levels
Reject rates decrease through increasing RFT (right first time)
Product cycle time decreased

Benefits specification

Potential benefit	Benefit metric	Benefit metric baseline	Accountability	Benefit metric target	Area of activity
Line 1 capacity	Packs/month	35	Line 1 Team Leader	70	Line 1 kaizen
Line 2 capacity	Packs/month	25	Line 2 Team Leader	50	Line 2 kaizen
Line 3 capacity	Packs/month	30	Line 3 Team Leader	80	Line 3 kaizen

(continued)

Table 2-19 (Continued)

Planning Toolkit – Benefits Realization Plan					
Project: Capacity Improvement Project			**Project Manager:** Jane Jones		
Date: February 15th			**Page:** 2 of 2		
Benefits specification					
Potential benefit	**Benefit metric**	**Benefit metric baseline**	**Accountability**	**Benefit metric target**	**Area of activity**
Product quality	% rejects	25%	Production Manager	5%	SOP design
Operating cost	Cost/capacity	$10000/pack	Production Manager	$10000/pack	Line modifications and SOPs
Cycle time	Batch time	120 days	Production Manager	90 days	All areas
Additional benefits may become apparent during the kaizen workshops					

Benefits realization				
Benefit metric	**Tracking frequency**	**Dependency**		**Priority vs vision of success**
		Project scope	**Business change**	
Line capacity	Monitor during and after project	Process observations and kaizen workshop	Line performance management (measures and team accountability)	HIGH
Product quality				HIGH
Cycle time				HIGH
Operational costs	Forecast as design progresses and then check after implementation	SOP design Equipment upgrades (parts)	Order placement and warehouse despatch SOPs	

Business change management

Achieving agreement of the Project Charter (Table 2-17) was seen as a major step in this project. A review of the current situation had already demonstrated that none of the three production lines was operating at or near their design capacity and that if they did, there would be no need for any additional equipment or personnel. However, taking a production process from a yield of 46% to 92% was seen as an impossible goal. Convincing senior stakeholders and then convincing those who are at the heart of the change process are two different things, and the Project Manager recognized that formal approval was only the start of 'unfreezing' mindsets about how this particular manufacturing facility was operated.

The operational team have clearly stated their objection to any improvement project on the basis that they understand the constraints of the production lines.

➤ They see an improvement project as a direct comment on their ability to deliver product to the customer, on time and within specification.
➤ They have had to deal with older equipment with little maintenance support for years and see a new line as the way to solve all their problems.

From a business perspective the installation of a new production line will require significant capital investment and take about 6 months to complete (mainly due to procurement timelines for the various equipment systems in the production line).

Therefore in order to get the operational team (who were to be the Project Team) fully engaged, the Project Manager had to understand their resistance and develop strategies to remove it.

The Project Manager got the teams together during one of the breaks (when the lines were all shut down) and asked them about the ways things were at the moment.

Based on the analysis of resistance (Figure 2-21) the Project Manager was able to strengthen the drivers for change by involving the operators at each stage and getting them to fully own solutions they had thought of.

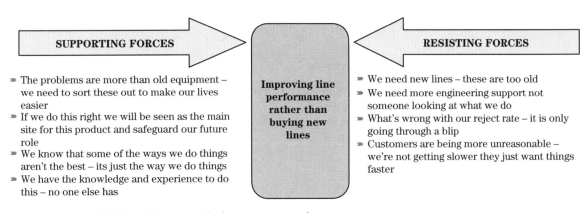

Figure 2-21 Force-field analysis – capacity improvement project

Conclusions

At the end of this project capacity was increased without requiring the installation of an additional production line and the associated resources to operate and manage it. Each of the existing lines was able to sustainably deliver the required increased capacity which did occur approximately in line with forecasts (Figure 2-22).

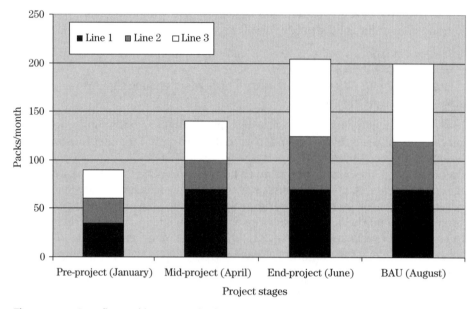

Figure 2-22 Benefits tracking – capacity improvement project

This required a major business change in terms of:

➤ Understanding the business processes required to deliver the product 'right first time'.
➤ Determining the root causes of the low yield.
➤ Operating the supply chain differently to focus on bottleneck management, 'right first time' production and empowered production teams.

Key points

The aim of using this particular case study was to demonstrate that:

➤ Even the most obvious business case can benefit from a review from a new perspective.
➤ There are many ways to deliver a set of benefits. The business plan is simply there to set out the selected way in which these benefits will be delivered once all options have been objectively considered.

Handy hints

Business planning is about the planning of good relationships between the project and the business

It is easy to forget that this element of planning the project is about relationships and not just planning the delivery of a business case. Project management must effectively integrate both 'hard' and 'soft' elements if it is to be robust and effective.

Sponsors need guidance

The majority of senior management who are assigned as sponsors will have had a varied experience of sponsoring projects. A good Project Manager will recognize the importance of guiding the sponsor and being very clear about what is needed from him.

Customers need management

Although we want to have a satisfied customer at the end of the project, we must also be clear on the difference between customer 'needs' and customer 'wants'. The management of customer expectations is an important activity and at a high level the sponsor can provide support.

Stakeholders need to be communicated with

The management of stakeholders is really about the management of communication – what you tell them, when and for what purpose. Recognize the importance of this element of the business plan and spend time working with the sponsor on the best way to deal with each stakeholder group and who is the main contact.

Although the majority of the business plan is only delivered at the end of the project (the benefits) it needs to be planned at the start

If you only start to consider handover, sustainability and benefits tracking when the project scope is delivered and handed over, then the chances of meeting the agreed business case are low. It is not easy getting people to plan at the best of times, so looking so far ahead is something you'll just have to stick at!

Know WHY you are doing a project

The business plan is a great way for the Project Manager to really understand how the project will deliver an organizational need. In doing so you might even discover a better delivery method that decreases uncertainty and/or improves the chances of success.

And finally . . .

A robust project business plan:

- Articulates the business rationale for the project – proving the organization needs it.

- Has a link to the project set-up plan – and its link to business benefits.

- Has a link to the project control plan – and its potential to be delivered in control.

- Is built on sound relationships and effective stakeholder management.

- Projects take a finite length of time and the world will be a different place by the time you have finished – the business plan needs to be able to cope with this level of uncertainty.

3 Set-up planning

In the context of project delivery planning, set-up planning is the way that we define how the project is to be administered. It is based on the approved project business case which has, or will be, developed into a business plan (Chapter 2).

The set-up plan supports the development and management of the relationship between the Project Manager and the Project Team. It is this relationship which needs to be planned. A Project Manager and Project Team who have a shared goal are more likely to deliver a successful project outcome.

The project delivery plan (PDP) must provide a robust methodology for the delivery of the project scope which will enable the realization of the business benefits, the delivery of which proves that the business case has been met.

Historically this part of a PDP has been reasonably well developed because it relies on the Project Manager being 'inward looking', towards the project. However, as described in Figure 1-6, there are two planning critical success factors (CSFs) which feed into the development of a set-up plan (which is a CSF itself):

- CSF 2: Engaged stakeholders – the stakeholders in the set-up plan are those involved in the delivery of the project: the Project Team.
- CSF 3: Capable Project Manager – the first stage in having a successful project organization is in selecting an appropriate Project Manager.

All projects need:

- A Project Manager and a Project Team with a clear, and shared, understanding of the project goal – a successful project outcome.
- A clearly defined roadmap for the team to follow, enabling clear planning and controlled delivery.
- A clearly articulated scope which enables the delivery of the business benefits and is a robust foundation upon which to control the project.
- A set of resources (people, money, assets) to enable and support project delivery.

Projects can fail when these essential roles, relationships and strategies are missing or are not effectively managed.

What is a project set-up plan?

Within a PDP, a project set-up plan is the formal articulation of 'how' the project will be administered so that the business is assured of certainty of outcome with respect to the original business case. It covers the following four planning themes:

- Project organization.
- Project type.
- Project scope.
- Project funding strategy.

This part of the PDP defines 'what' is to be delivered by the project, 'who' will be involved in the delivery and 'how' resources are to be engaged. It defines the overall project roadmap, the route the project will take during its delivery.

How to develop a project set-up plan?

The Project Manager would typically rely on a core part of his Project Team in order to develop the set-up plan. The focus of this type of planning is the ability to deliver the defined scope through engaging the required capability and capacity of personnel.

To develop an effective set-up plan a Project Manager needs to fully understand the scope needs of the project and this inevitably requires team members' involvement and engagement.

At this stage a Project Manager will have been assigned to the project and the appropriateness of this selection is crucial in developing an effective set-up plan.

Selecting a Project Manager

The selection of a Project Manager is crucial to both project delivery planning success and overall project success. For major projects within an organization it is not uncommon for the PDP to be developed by a team of Project Managers with only one being assigned for the actual delivery. This strategy recognizes the criticality of planning to ultimate project success.

When an organization makes a commitment to make a change to the business through the delivery of a project, the selection of the Project Manager must match the needs of the project. These are usually described in terms of

- Project type.
- Project size (exact meaning of 'size' depends on the project type).
- Project complexity.
- Criticality of the project to the business.

A typical selection model is shown in Figure 3-1. This uses all the above criteria with the exception of project type. This parameter is usually considered when size, complexity and criticality have been determined.

The model supports an organization in developing project management (PM) capability as it recognizes that Project Managers need to develop through a career path from novice to expert. In addition, it highlights the level of project management capability needed to meet different project challenges: size, complexity and criticality.

		Small	Medium	Large
Project complexity/criticality	High	**Competent to high PM capability** High experience of the project type	**High PM capability** Experience of medium complexity and medium size on this project type	**Expert PM capability** Likely to have handled either similar size OR similar complexity previously for this project type
	Medium	**Developing PM capability** Experience of the project type	**Competent PM capability** Experience of the project type	**High PM capability** Experience of a variety of project types Experience of similar size of project
	Low	**Novice PM capability** Technical understanding of the project type	**Developing PM capability** Experience of this type and size of project	**Competent to high PM capability** Some experience of the project type High experience of medium-sized projects

Project size

Figure 3-1 Selecting a Project Manager

Project type

Many organizations place Project Managers in 'silos', believing that they can only deliver a specific project type. Whilst this is sometimes the case there is significant evidence that the foundation of good project management practice is generic, the principles being the same no matter the project type.

Project size

Many organizations consider financial value as the unit of project size, particularly when dealing with projects funded from capital rather than expense. Others look at team size, project duration and/or benefits targets. Each of these parameters is linked to scale of activity and in effect it is this which we assess when considering selection of personnel. It is difficult to move from managing a 5-person Project Team to a 50-person Project Team, for example, or from managing a capital budget of $100,000 to $100 million.

Project complexity

Complex projects are a greater challenge at all stages, however, risks can be greatly mitigated by thorough planning. Complexity is usually linked to the dependencies within a project: activity, deliverables (outputs), resources, benefits or the dependencies to things external to the project (such as access to assets or information). Complexity is independent of project size: it does not necessarily follow that a large project is complex or vice versa.

Project criticality

The importance of the project outcome to the organization is a different parameter from complexity, yet it is usually coupled with it. A critical project may be of either low or high complexity; however, in terms of selecting a Project Manager an organization is likely to want higher capability independent of complexity. An organization is looking for certainty of successful outcome for a project deemed critical to its business operations and/or future sustainability.

Short case study

A small organization was proposing to invest significant capital over the next 3 years. This was to be managed as a portfolio of capital projects ranging from equipment replacements to full re-builds. A number of the smaller-sized projects were extremely simple (specify and buy equipment) whereas one or two were large, complex and critical to the business operations (change areas of the supply chain, introduce new products and improve quality).

Initially the finance director maintained overall control of the capital funds and assigned resources to each individual project from the pool of experienced corporate Project Managers. However for the later projects he had to rely on site based Project Managers with less experience of complex or large projects.

Nine months into the 3-year programme the CEO reviewed progress and realized that the selection of Project Managers needed to be changed in order to manage some of the issues which had arisen:

➤ Inappropriate selection of Project Managers: assigning experienced ones to simple projects and less experienced ones to major projects.
➤ No management of the dependencies between the various projects (e.g. resources, scope, assets).

The projects were all put on hold for 3 weeks, and an experienced Project Manager was assigned to the overall portfolio, allowing him to reassign his team in order to minimize risks and maximize potential for success.

Project management capability model

A question often asked by project management professionals is 'are good project managers born or made?' This leads on to the following questions:

➤ Can anybody be a good Project Manager if they get the 'proper' training?
➤ Can you train someone to be focused on the delivery of a set of agreed project objectives?
➤ Are tools and techniques at the core of what makes a good Project Manager?

Whilst these questions and more could be answered here from the authors perspective it is more useful to consider the general theme of project management capability.

Definitions

Table 3-1 defines the main terms used in the project management capability model which is summarized in Figure 3-2.

Table 3-1 Project management capability model definitions

Term	Definition (1)	Definition (2)
Capability	Our ability to perform	A function of competency and performance (current and past)
Competency	Our ability potential	A function of a person's knowledge, skill and behaviours
Knowledge	The things we know	A person's awareness and understanding of facts
Skills	The things we can do	A person's practical ability to apply knowledge
Behaviours	The way we act	A function of our personality, values and attitudes in a particular environment or culture
Performance	The way experiences have progressed	The act of doing something successfully

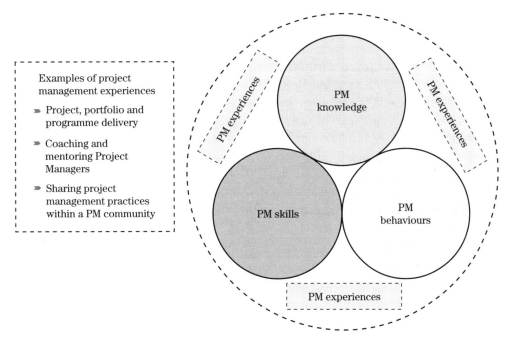

Figure 3-2 A project management capability model

All elements of capability can be measured and most by a suite of tools which give both qualitative and quantitative results. All elements of capability can be improved to meet a developmental goal:

➤ Knowledge can be built through learning.
➤ Skills can be learnt through practice.
➤ Behaviours can be changed through changing attitudes or values.
➤ Performance can be enhanced through new experiences.

Figure 3-2 takes the generic definitions and applies them to project management. It proposes that project management capability can be assessed by measuring a person's PM knowledge of the subject, their skill in using PM tools and processes and their behaviours during the management of projects as well as their delivery success.

Project Manager capability profiles

Based on the project management capability model (Figure 3-2) techniques have been developed to quantify a Project Manager's capabilities in all areas. Figure 3-3 shows one simple example of this.

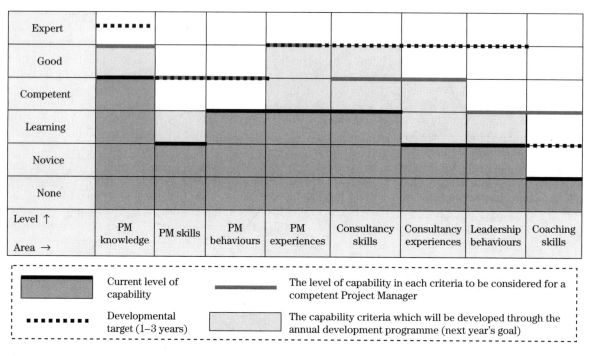

Figure 3-3 Example project management capability profile

Here the Project Manager is seen to be at a learning level with a 3-year goal of being competent. In the next annual development cycle he has selected areas to work on and in this case has highlighted project and consultancy experiences as major developmental goals, alongside development of the required organizational leadership behaviours.

Project management knowledge

A Project Manager's awareness and understanding of principles, processes and tools can be measured through use of standard training courses and links to specific bodies of knowledge. This can be quantified by:

➤ Checking what we know versus standard project management bodies of knowledge.
➤ Completed recognized training courses aimed at increasing awareness and understanding of project management knowledge.

Project management skills

A Project Manager's skill in using relevant knowledge through selecting and using the appropriate tool or process for a given situation can be measured during projects. This can be quantified by:

➤ Checking the project management tools and processes used within a variety of projects (Table 3-2).

Table 3-2 Sample project management toolkit assessment

Toolkit		Example tools	Skill level (%)	Overall skill level (%)
Basic	Stage one	'Why?' Checklist	53	
	Stage two	'How?' Checklist Project Gantt Chart RACI Chart	64	
	Stage three	'In control?' Checklist Action logs	80	67
	Stage four	'Benefits realized?' Checklist Project after action review	40	
Standard	Stage one	Business Case Tool Benefits Hierarchy	16	
	Stage two	Project Charter Control Specification Table Logic linked Gantt or PERT Charts	60	29
	Stage three	Milestone Chart Risk assessment and risk logs	20	
	Stage four	Benefits Tracking Tool	8	
Expert	Stage one	Benefits mapping	10	
	Stage two	Contract selection matrices Resource loaded Gantt Charts	16	8
	Stage three	Earned Value Tool Schedule contingency rundown	8	
	Stage four	Benefits Scorecard	0	

➤ Completed recognized training courses aimed at applying knowledge in a controlled, simulated project environment.

The example toolkit assessment (Table 3-2) is taken from a more detailed toolkit where each tool within a stage of the project has been listed. The user then states whether the tool has been used and at what level. This assesses both knowledge and skill but it is primarily intended to track the development of project management skills. In the example the user has a good basic project management skill level and is developing a standard skill level. The data also shows that the user is developing a strong stage two (project delivery planning) toolkit with much less experience in stage four (benefits realization).

The full spreadsheet is contained in Appendix 9-16. This can be adapted for use within specific organizations and is also a useful tool to assess project management competence within a Project Team.

Project management behaviours

There are a specific set of identified behaviours which are proven to be an indicator of good project management such as strategic thinking, getting things done, focus, personal drive and energy. These can be quantified by:

➤ Measurement of feedback from third party observations such as those used during 360-degree performance assessment processes.
➤ Checking the positive and less positive behaviours observed on recent projects and comparing to a standard profile (Figure 3-4).

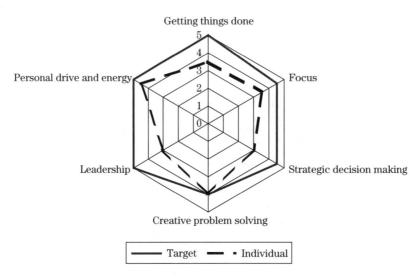

Figure 3-4 Example project management behaviours assessment

A Behaviours Assessment Tool which can allow an individual Project Manager to compare his observed behaviours with those of a best practice benchmark can be used as a developmental tool. In addition, it can be used to review whether the Project Manager's profile 'fits' with the particular project challenge. For example, a Project Manager with the profile shown in Figure 3-4 might find a large, complex programme of projects difficult to manage due to the gaps in strategic thinking and leadership. In these types of projects the Project Manager may be managing a core team of Project Managers and taking a more strategic, leadership role.

There has been a lot of research conducted into how individuals behave if they are 'naturally' good at a particular role. The majority of models work on the basis that most people have:

- Inherent abilities due to their personality, values and natural abilities: nature as opposed to nurture. These are not 'trainable'.
- Developing abilities due to their attitudes and the environment and culture in which they operate: nurture as opposed to nature. These are not 'trainable' but develop with life experiences in various environments.

Project management experiences

It is vitally important that a Project Manager uses his developing skills and knowledge on 'live' projects. The history of these experiences can be quantified by:

- Use of a project scorecard to track each project delivered, using key performance measures to benchmark performance of a specific project against best practice. Example metrics would typically be:
  - Percentage schedule adherence.
  - Percentage cost adherence.
  - Delivery of business case – red, amber or green.
  - Customer satisfaction – score rating 1 (low) to 4 (high satisfaction).
- Use of a project history document (Figure 3-5) to summarize a Project Manager's highs and lows; greatest achievements and most significant challenges:
  - Focus on breadth and depth of experiences within each different project type.
  - Focus on generating a balanced set of experiences which supports development of skill, knowledge and behaviours, as well as improving project performance.

Project sustainability performance	Project management processes
Assessment against business criteria: benefits realized and changes sustained for customers	Assessment of project management processes used; their effectiveness and experiences in different project situations
ACHIEVING CUSTOMER SATISFACTION	**ACHIEVING EXCELLENCE IN PROJECT MANAGEMENT**
Personal learning and development	Project delivery performance
Assessment of learning from project experiences against current and future development goals: consider 'gaps' seen and particular challenges	Assessment against tangible measures: deliverables or meeting cost or timeline commitments
ACHIEVING PERSONAL GROWTH	**ACHIEVING BUSINESS GOALS**

The aim of any scorecard tool is to present a balanced view of performance

The theory of the balanced scorecard is to measure performance within four areas to support 'balance' within the organization

- Financial
- Customer
- Internal processes
- Learning and growing

A Project Manager scorecard uses objective data from project experiences to consider:

- Project delivery performance
- Project sustainability performance
- Project management processes
- Areas of personal learning and development

The scorecard assumes that for every project experience there are both personal and business benefits and that there needs to be an appropriate 'balance' between these

Figure 3-5 Example Project Manager scorecard

Additional criteria

Project Managers need to have other more traditional management skills if they are to be successful. Most capability models' allow for additional criteria linked to leadership, people, change management and consultancy skills and experiences.

Summary

This type of detailed project management capability model can support an organization in defining any cumulative project management capability gap, as well as support Project Manager selection for a specific project or a general recruitment need.

Tool: Project Manager Selection Checklist

Based on usual selection criteria (pages 75 and 76) a this tool has been developed to check that the selected Project Manager is the right match for the specific project (Table 3-4).

Table 3-4 The Project Manager Selection Checklist explained

Planning Toolkit – Project Manager Selection Checklist		
Project: *<insert project title>*	**Sponsor:** *<insert name>*	
Date: *<insert date>*	**Project Manager:** *<insert name>*	
Project type		
Type of project *<insert the type of project: capital or expense, engineering, IT, manufacturing or business process. Use whatever parameter is appropriate based on the main focus of the project>* **Applicable project management experiences** *<insert all project management experiences on similar projects and state the formal role on these>*		
Project size		
Size of project *<insert the size of project using the most appropriate criteria: financial value, anticipated duration, team size, scale of activity>* **Applicable project management experiences** *<insert all project management experiences on similar size projects and state the formal role on these>*		
Project complexity/criticality		
Project complexity *<insert the level of complexity and then expand on the areas which are of medium and high complexity>* **Criticality of the project to the business** *<insert whether this project is low, medium or high priority for the business in terms of its short, medium or long term goals, outline the worst case scenario for the business if the project failed>* **Applicable project management experiences** *<insert all project management experiences on similar complex and/or critical projects and state the formal role on these>*		

(continued)

Table 3-4 (Continued)

Planning Toolkit – Project Manager Selection Checklist			
Project:	*<insert project title>*	**Sponsor:**	*<insert name>*
Date:	*<insert date>*	**Project Manager:**	*<insert name>*
Capability development			
Capability gap *<insert any area where there is a lack of capability and therefore a potential risk>* **Capability development goal** *<insert how identified capability gaps are to be managed so that the gap has been eliminated by the end of the project, confirm any gaps which cannot be managed through development whilst delivering the project>* **Required support needs** *<insert any support needs for the Project Manager in order to meet the developmental goals: mentoring, additional project reviews, overview project management by another>*			
Decision			
Selected Project Manager confirmed *<select Yes/No, YES means that all support needs will be made available>* **Additional comments** *<insert reasons why the Project Manager is or is not the right match for this project>*			

The tool is also used to:

➣ Confirm the development capability goal for the Project Manager.
➣ Define any additional resources the Project Manager needs in order to be successful.

This checklist would be used by those in the organization responsible for project management and therefore the development of project management capability. They are also usually senior stakeholders in terms of project success.

Project organization

The heart of any project is the people who are tasked with delivering the project scope so that the business receives the anticipated benefits. Following selection of the Project Manager, an organization has to consider what internal resources it has available to support the project on either a full or part-time basis. This selection process is the first stage in ensuring that the appropriate level of capability is integrated into the Project Team. The next stage usually involves looking for capability external to the organization (Chapter 4). The set-up plan therefore has to integrate this internal 'people plan', covering:

➤ Team selection and capability development.
➤ Team start-up.
➤ Team performance planning.

Team selection and capability development

Having the right mix of skills and experience in a Project Team is a key part of being able to deliver success. Often teams are pulled together through a variety of inappropriate approaches and this can lead to problems before the project has even started:

➤ Project Teams made up of the only people available often have the incorrect capability profile to deliver the project, with key gaps needing to be filled through external sources.
➤ Project Team structure is typically hierarchical and also functionally based, leading to a 'silo' mentality with in a project. Team members may think only of their own functional targets rather than those of the overall project.

Team capability profile

A sound methodology for selecting a Project Team (or to evaluate assigned team members after less robust selection processes) is to develop the necessary team capability profile for project success. Figure 3-6 shows an example of how this can be used to evaluate risks associated with team selection.

A Project Manager has to initially determine the key criteria for performance success and then determine an objective process for evaluating potential team members. The selection process must then look at the total set of capabilities within the team and decide if it matches the requirements.

The example in Figure 3-6 shows the evaluation of a team of sub-project leaders within a large project. The Project Manager determined that as long as each team leader had one capability in excess of a score of 80 with one other over 70 then the profile was adequate as long as at least one team leader was an 'expert' in each of the four capability areas.

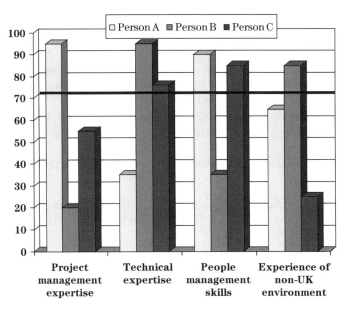

In this example a Project Manager is reviewing the capabilities of the assigned team leaders in his project

Persons A, B and C all have a role in before review leading a substantial sub-project within a major global logistics project

The Project Manager recognizes that a score of 70 or more on each of these criteria would give him a strong team with minimal risks

Each person has a gap and the Project Manager has to consider the risks of that gap on the project as well as any potential strategies for eliminating the gap or mitigating its effect

Figure 3-6 Example team capability profile

Another way to evaluate this is shown in Figure 3-7. In this example the company has developed a very typical organization chart for an engineering project (described in Figure 3-8, before review):

- The engineering disciplines have been segmented into functional teams.
- Each team has a functional team leader and functional resources.

Skills required → / Available resource ↓	Mechanical engineering and design	Project management	Computer systems validation	Automation engineering	Civil and structural engineering and design	
Name A	✓				✓	⎫
Name B	✓	✓	✓			⎬ Selected team members
Name C			✓	✓		
Name D	✓	✓			✓	
Name E		✓	✓	✓		⎭
Name F					✓	
Name G	✓	✓				

Figure 3-7 Example skills matrix

The Project Manager highlighted five key areas of engineering experience which were needed within the project and reviewed each assigned team member against them. As a result he made the following changes:

➤ Removal of person F and G from the team on the basis that the skills they had to offer were already within the team (resource capacity needs were also checked before this decision was made).

➤ The civil engineer was given overall authority for the delivery of the mechanical, civil and structural scope with additional technical expertise being available from the Project Manager.

➤ The mechanical and instrument engineers were considered a design team and were required to work together to develop overall plans which integrated all civil, structural, mechanical and automation needs.

➤ The Project Manager negotiated more direct access with the customer so that his expertise could be called upon as the design progressed.

Figure 3-8 shows the before and after organization charts. Not only did this release resources back to the company (and therefore reduce project costs) but it changed the culture of the Project Team: from traditional functional into cross-functional.

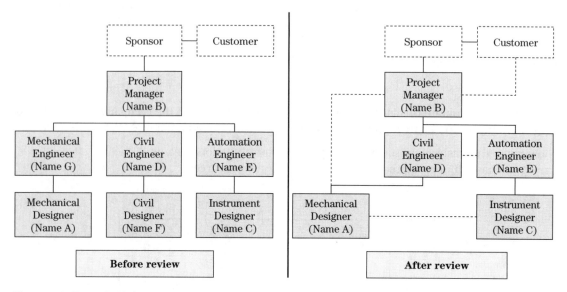

Figure 3-8 Example Project Team organization charts

In addition Figure 3-8 shows how a simple organization chart can be used to visually display relationships other than just management links.

Team structure

Figure 3-8 also demonstrates the very traditional nature of project organization design. Structures are typically hierarchical and functional. Project Managers need to understand the variety of team structures available to them and to select the most appropriate one based on specific project needs and challenges (Table 3-4).

Table 3-4 Project Team structures

Structure	Description	Pros	Cons
Functional	Specialist skills are segmented into sub-teams lead by a specialist team leader	➤ Clear reporting lines ➤ Collation and development of specialist skills ➤ Specialist innovation encouraged ➤ Works within a strong centralized project culture and structure ➤ Allows efficient use of specialist resources	➤ Can get very insular and not consider wider implications of specialist solutions ➤ Functional goals can become misaligned from project goals ➤ Can limit horizontal communication within the project
Project outcomes	Sub-teams are developed from consideration of project outcomes	➤ Builds a strong sub-team identification with the area of project scope ➤ Strong cross-functional working which encourages multidiscipline working ➤ Clear performance targets can be set linked to overall project goals ➤ Can operate in a more flexible project organizational culture with devolved responsibilities and an empowered culture	➤ Requires functional resources in each sub-team which can limit resource levelling and therefore overall efficiency ➤ Can still retain a very vertical, hierarchical organization ➤ Coordination between sub-projects can be limited and this can be a bigger issue if the dependencies between the sub-projects are not managed well
Geography	Sub teams are determined by the geographic location of either resources or where the project is being delivered	➤ Builds a strong team identity based on the location ➤ Strong cross-functional working which encourages multidiscipline working ➤ Clear performance targets can be set linked to overall project goals	➤ Location related goals can become misaligned from project goals ➤ Can still retain a very vertical, hierarchical organization ➤ Requires functional resources at each location (which may not always be available)
Matrix	Sub-teams have dual reporting lines which can relate to any of the above	➤ Build strong links both functionally and scope or location related ➤ Can operate in a more flexible project organizational culture with devolved responsibilities and an empowered culture ➤ Encourages strong level of coordination across both reporting lines ➤ Requires a strong decision-making process to align with the dual authority structure	➤ Typically one link needs to be major and one minor or issues can occur ➤ Resources tend to be allocated to multiple teams in order to be fully utilized ➤ Can develop some confusion regarding role, responsibilities and performance targets through having two reporting lines ➤ Budget management can become more complex (particularly expense costs) ➤ Increased coordination can increase administration and management costs

A Project Manager would typically review the key information flows necessary to deliver the project scope successfully, review team capability needs and then design the final structure once the team members have been assigned.

Only once the structure has been developed can roles, responsibilities and performance targets be allocated. In addition the organization chart can be used to indicate broader working relationships and information flows. This adaptation of an organization chart is usually referred to as a relationship map (Figure 3-9).

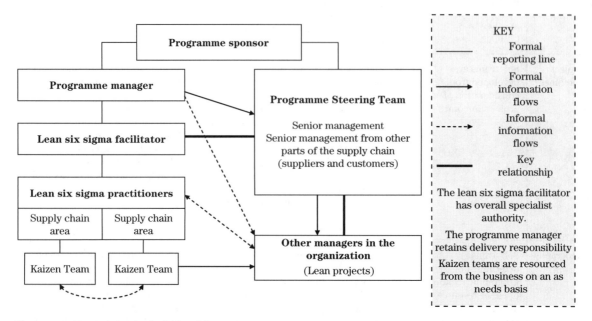

Figure 3-9 Example project relationship map

Tool: Team Selection Matrix

The aim of this tool (Table 3-5) is to support the articulation of project capability needs through the definition of key capability criteria and then the use of those criteria to select appropriate Project Team members.

Table 3-5 Team Selection Matrix explained

Planning Toolkit – Team Selection Matrix						
Project: *<insert project title>*			**Sponsor:** *<insert name>*			
Date: *<insert date>*			**Project Manager:** *<insert name>*			
Potential team member	**Selection (yes/no)**	**Selection criteria**				
		Criteria 1	**Criteria 2**	**Criteria 3**	**Criteria 4**	**Criteria 5**
<insert name of person under consideration>	*<insert decision>*	*<insert individual score vs criteria>*				
<insert name of person under consideration>	*<insert decision>*					
Summary						
<insert any summary comments regarding the final selection and also any gaps which still remain, how these may be eliminated and any risks which need to be managed within the team>						

Potential team member

At some stage in the selection process there will be a list of potential candidates who need to be formally assessed. These people may well have already gone through a first filter depending on whether the selection is for a specific sub-team or a general capability review of the whole team.

For example in a large project decisions may already have been made about team structure based on the scope needs. This may drive a different set of capability needs for each sub-project and therefore the potential candidates may be drawn from a large pool of available resources. To avoid having to assess each candidate against each separate sub-project need, a Project Manager may use one filtering criteria, for example:

'Sub-project team 1 will be based on site in South America and so only resources willing to relocate for the 12-month construction period can be considered.'

Selection

A clear decision needs to be recorded. Any explanation of the decision can be put into the summary section if necessary. This is often needed if senior stakeholders have been championing a specific candidate.

Selection criteria

The criteria should be based on some element of capability: knowledge, skills, behaviours and/or experiences (proven performance). They should be able to be used to objectively assess a candidate and be capable of allowing comparison between candidates. For example, selection of an engineering team may use 'be a chartered engineer' as a filtering criteria rather than a more specific selection criteria such as 'have experience of current safety legislation on chemicals facilities'. In this way the engineers can have length and type of experience compared.

Summary

The Project Manager would use this section to highlight gaps and current risks and how these are to be managed so that the project has the highest potential for success.

Team start-up

The goal of any Project Manager is to have an effective and high performing team. This requires planning and significant work prior to delivery. A Project Manager, therefore, needs to have an understanding of how teams become high performing.

High performing teams

It takes time, energy and planned effort to get a team to be high performing. For some activities in the workplace it is appropriate for people to work as individuals in a group. However, in a project, success is linked to the combined effort of a team of people with specific capabilities and so high team performance is required. Table 3-6 compares the attributes of a group and various types of teams based on three variables:

- Team effectiveness
- Team trust
- Team performance

Table 3-6 Teams versus groups

	Team Effectiveness	Team Trust	Performance
Group	➤ Very low ➤ No collective outputs	➤ Low to medium ➤ Respect for individuals in the group	➤ The group achieve its goals through achieving individual goals
Nominal team	➤ Low to medium ➤ Collective outputs are generated and are generally acceptable	➤ Very low ➤ Team members are not prepared to take any risks	➤ The individuals are a team in name only and performance is low
Newly formed team	➤ Medium ➤ Collective outputs are generated and are acceptable	➤ Low to medium ➤ Respect for individuals in the team	➤ Medium ➤ The team achieves its goals
Average performing team	➤ Medium ➤ There is commitment to team success	➤ Medium ➤ Team members are starting to trust each other	➤ Medium ➤ The team achieves its goals
High performing team	➤ High ➤ There is a strong commitment to team success through supporting each other ➤ The team work highly effectively as a unit to deliver outputs	➤ High ➤ Team members have a high degree of trust in each other and can take risks (such as positive conflict)	➤ High ➤ The team exceeds its goals through delivering requirements and other 'intangibles'

High performance teams do not just 'happen'. They require time and effort in order to build them appropriately. If a Project Manager wants to achieve this goal then there are specific activities he must do to support appropriate team behaviours.

Team building

Teams do not build themselves. They need leadership and a purpose. They too need to analyse what makes them tick, both as individuals and as a team. The following model (Table 3-7) describes the stages a team would typically go through.

Table 3-7 Team building model

Stage	Description	Team behaviour	Project Manager role
Forming	The team is brought together	<li>There is both enthusiasm and a little trepidation as all team members are facing a new challenge</li><li>Team members are more concerned with individual rather than team goals</li><li>Conflict avoidance</li>	<li>Communicate a clear vision of project success and translate this so that all team members can see their role in it</li><li>Work with the team to set clear ground rules for behaviour and actions</li><li>Be consistent in management style so that the team can understand to what extent they are empowered</li><li>Contract with each team member</li>
	Suggested activities: Hold a team launch event to introduce team members to each other and to the challenge they are facing as a team		
Storming	The team enters a chaotic stage	<li>Team members have differences of opinion and express them</li><li>Vying for position and role</li><li>Assumptions, roles and objectives are challenged</li>	<li>Recognize that conflict can be constructive nut needs to be managed</li><li>Don't try to stifle conflict unless it gets personal rather than about the project</li><li>Work with the team to highlight individual strengths and weaknesses</li>
	Suggested activities: Hold team meetings which are based around brainstorming activities to allow all team members the freedom to express their individual opinions		
Norming	The team begin to work together	<li>Team members generally agree with each other or can reach consensus</li><li>Team processes are developing and individuals have clarity</li>	<li>Recognize that at this stage the team are still only the sum of its parts and needs to be managed to the next stage of team development</li><li>Reinforce the team goals as well as individual objectives and establish a culture of participation</li>
	Suggested activities: Hold team meetings focused on measuring performance		
Performing	The team are performing effectively	<li>Team members see the team as a safe place to work where differences are welcomed constructively</li><li>There is value and respect for each other</li>	<li>Recognize the team is a highly effective unit and their contribution is greater than the sum of the parts</li><li>Manage the unit with an appropriate level of empowerment establishing a culture of delegation as appropriate</li>
	Suggested activities: In team meetings encourage synergistic activities, creativity and innovation whilst also continuing to focus on delivery		
Mourning	The team is dissolving	<li>Team members are starting to think of the next role (as individuals)</li><li>Goals have been achieved</li>	<li>Management of the project to 100% completion</li><li>Recognize team achievements and also specific individual achievements, give feedback</li><li>Disengage with each team member</li>
	Suggested activities: Hold a team after action review (AAR) to allow members to recognize value they have delivered, celebrate success and to say goodbye		

The interesting thing about team dynamics is that any change can cause the team to move to a previous state. For example, if a key team member leaves before the end of the project the team is thrown into mourning and then has to reform once a new team member joins. This inevitably means that storming and norming will once again occur on the route to effective performance.

During team development the Project Manager and team members should be using the consultancy process (page 16), contracting with each other and building one-on-one relationships.

In addition the Project Manager would be using the continuum of empowerment (page 227) to support definition and articulation of the most appropriate level of empowerment.

Team communication

During team start-up and management the most important process is that of team communication. Effective communication supports team success; however, within a project there are many potential gaps which cause either complete project failure or a reduction in project performance (Figure 3-10).

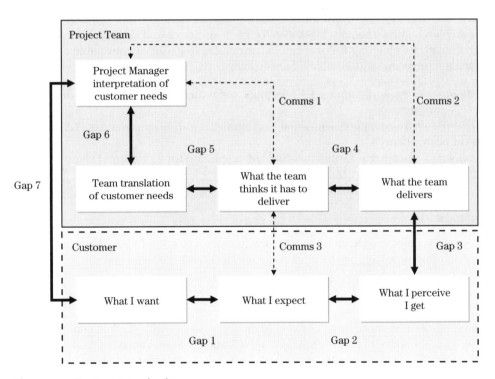

Figure 3-10 Team communication

Key to Figure 3-10

➤ *Gap 1* – Customer expectations are low based on previous experiences.
➤ *Gap 2* – Customers sometimes don't know what to expect and this can lead to a perception gap – 'I didn't think it would look like that'.

- *Gap 3* – The customer disappointment gap.
- *Gap 4* – This is a team performance gap linked to the lack of aligned cascade of scope (quality, quantity and functionality) throughout the team.
- *Gap 5* – This is a team performance gap linked to the lack of team effectiveness.
- *Gap 6* – The team generates a detailed specification which does not match what the Project Manager has interpreted as the customer needs. This is a team management gap.
- *Gap 7* – This is an interpretation gap. The Project Manager did not fully understand what the customer wanted.
- *Comms 1* – A communications route to mitigate gaps 5 and 6.
- *Comms 2* – A communications route to mitigate gap 4.
- *Comms 3* – A communications route to mitigate gaps 1, 2, 3 and 7.

Communication vehicles

Although there are many more communications routes, the three shown in Figure 3-10 are critical team communications routes – between each other, with the Project Manager and with the customer. Each route may have a different vehicle and the role of the Project Manager is to develop an internal communication plan highlighting key messages for each audience and the selected vehicle. Internal project communications planning follows similar principles to external communications planning (page 35). Typical communications vehicles are:

- *Project Manager to team member* – 1-1 meetings, team meetings, project bulletins, project notice boards, project reports.
- *Team member to team member* – informal get togethers, sub-team meetings, 1-1 meetings, action lists, project notice boards.
- *Team member to customer* – formal meetings on a specific topic, informal chats as a part of customer engagement, project bulletins.

Team processes

During team start-up there are various processes which need to be defined. These generally support the team in understanding how they are to work together to deliver the team goals.

Ways of working (WoW)

In general the process of working together as a team needs to be defined. This usually covers issues such as team location, team area set-up (both virtual and non-virtual), ground rules and role definition.

Project teams can be located in various offices or geographic locations, pulled together in one location as a task team or be completely virtual (using web, video and teleconferencing to communicate and work together).

Team roles may be defined as formal role specifications as well as less formal roles. For instance, many teams use Belbin team roles (Belbin, 2004) to identify team member characteristics. Knowing who is the 'ideas' person or who 'gets things done' is an important part of the development of team WoW and supports team development.

In an increasingly virtual and paperless world setting up the electronic systems and processes is crucial to effective start-up and subsequent team performance. Being able to quickly access people and/or documents allows teams to be agile and responsive.

Information flows

An important process which requires definition is how information will flow around a project. Team members need to be kept informed of some activities, be consulted on others and also understand who they should keep in the loop when they complete an activity. A very high-level way to do this is to use a RACI Chart (Table 3-8). Not only does this tool help to define responsibilities and accountabilities for individuals/roles, but it sets the high level flow of information.

Table 3-8 Example RACI Chart

Project Management Toolkit – RACI Chart					
Project: Business Change Implementation			**Date:** Week 2		
Names → **Activity ↓**	**Assignment lead**	**Project Team**	**Sponsor**	**Programme lead**	**End user**
Sponsor contracting and contract plan	A, R	I	C	I	–
Stakeholder engagement (team, end user and business area teams)	A, R See note 1	R See note 1	R See note 1	I	C
Business change implementation	A, R See note 1	R See note 1	R See note 1	I	R See note 1
Review implementation effectiveness	A, R	C	C	I	C
Implementation tracking	A, R	C	I	–	C
Completion of mandatory tools	A, R	C	I	I	C
Sponsor and end user review and approval	A, I See note 2	I	R	I	C
Project handover	A, I See note 3	C	C	I	R
Disengagement (including AAR)	A, R	C	C	I	C

Other tools used to describe information flows are flow charts and document matrices.

Decision making

Within any team the level of empowerment needs to be clear and apparent to all team members. Typically this means defining what decisions can be made by whom. For some technical decisions there will be a clear, and sometimes legal, precedent. For example a junior engineer has work approved by a senior engineer prior to release because within the organization the decision to release a technical specification has to be made by a chartered engineer.

Typical tools to define decision making include role profiles (Table 3-9), decision flow charts and decision matrices.

Training and development

During team selection some capability gaps may have been defined and/or development goals agreed. Therefore the management of team training and development is an important process to set up at the start of a project. Typically this would be done using a training matrix and individual training folders for all team members.

Tool: Project Team Role Profile

The aim of this tool (Table 3-9) is to support the definition of key roles within a Project Team so that individuals are aware of their own role outline as well as those of others. It is intended to collate the key roles to ensure alignment of objectives, clarity of decision-making authority and interactions necessary for a positive project outcome.

Table 3-9 Project Team Role Profile explained

Planning Toolkit – Project Team Role Profile			
Project: *<insert project title>*		**Sponsor:** *<insert name>*	
Date: *<insert date>*		**Project Manager:** *<insert name>*	
Role	**Description**	**Performance metrics**	**Key relationships**
<insert team or individual role>	*<insert main responsibilities – in terms of activities, decisions and deliverables>*	*<insert key measures of role success>*	*<insert key relationships which are necessary to achieve performance measures>*

It would usually be supplemented with a team RACI Chart and a team decision-making matrix or flowchart.

Role

The role title and current selected team member (if known) would be inserted. It is important that the role title matches that included within the project organization chart and other key documents such as RACI Charts.

Description

Although a role may be involved in many activities within the project (as described by the RACI Charts) it is important that only main activities, decisions and/or deliverables are listed here.

For example, an operator from a manufacturing facility may take on the role of a team leader within a manufacturing improvement project. Although he will be involved in most project activities, his main role is to deliver a new standard operating procedure (SOP) for the improved facility. In addition he is a part of the team which will make go/no go decisions on improvement options. It is important to describe the role fully but succinctly.

Performance metrics

The team will have a set of performance metrics; however, in terms of the management of an individual it would be usual to set performance metrics which align with the overall team goals as well as individual developmental needs. The metrics for an individual need to be set carefully in order to drive the correct behaviours.

For example an individual goal to complete an activity on time cannot override the need for the team to deliver its overall intent. The operator in the above example cannot deliver his SOP in isolation of the delivery of the improvements, nor can he select the improvement options which would make it easier for him to develop the SOP. As an engaged team member the operator wants the best decisions to be made even if this means that the SOP is delivered later than planned.

Key relationships

It is important to identify those relationships, either within or external to the team, that are an inherent part of the role. In our example, the operator is representing the manufacturing facility and needs to maintain appropriate links with business as usual (BAU): the operators and management team of the manufacturing facility.

Tool: Team Start-up Checklist

Sometimes teams can be formed quickly and a key step is missed in getting the team working well together. Once a Project Manager has started the project, it is often useful to check that the start-up of the team has been completed. The aim of the Team Start-up Checklist (Table 3-10) is to perform this audit and allow the Project Manager to put any required action plans in place early in the project life (prior to actual delivery).

Table 3-10 Team Start-up Checklist explained

Planning Toolkit – Team Start-up Checklist		
Project: <insert project title>	**Sponsor:** <insert name>	
Date: <insert date>	**Project Manager:** <insert name>	
Administration		
Have all selected team members been released for this project? <insert yes/no and attach a list of any resource release issues or concerns> **What start-up information has been issued to team members?** <insert list of information issued to individuals or to team leaders for cascade to team members> **Are team members aware of the location (physical or virtual) and their place in it?** <insert yes/no and attach information to support this>		
Launch event		
Has a kick-off meeting been held? <insert yes/no and attach the IPO or agenda for the meeting either held or planned> **What were the main outcomes of the kick-off event?** <insert actual or anticipated outcomes from the event in terms of actual deliverables to support team performance and management>		
Performance		
Has a vision of success been generated and shared with the team? <insert yes/no and attach vision and path of CSFs> **Have team ground rules been generated and shared?** <insert yes/no and attach a copy of the agreed ground rules developed and agreed by the team> **Are all team members clear about their role and responsibilities?** <insert yes/no and attach high level team RACI Chart and role profile summary> **Do all team members have clear performance targets aligned to project goals?** <insert yes/no and attach summary of performance metrics>		
Summary		
Has the team been effectively launched? <insert yes/no> **Are any actions required to support effective team working?** <insert yes/no and attach an action plan as appropriate and/or a summary risk assessment>		

Administration

Whether a team is part-time, full-time or virtual, organizing the logistics is important:

- Team meeting times and locations.
- Team working areas including access to phone, email and project information.
- Team virtual environments for sharing of project data.

Often the most infuriating part of being a part-time team member is not having appropriate access (either physically or virtually) to team areas or data and not knowing what is going on. In addition, lack of appropriate equipment can get in the way of effective team working.

Launch event

At some stage the selected team members will come together for the first time. This is the launch event. Most successful launch events have a secondary purpose other than just everyone meeting each other. Typically these are:

- Development of ground rules.
- Setting out the project goals.
- Developing the scope.
- Assigning roles and responsibilities.

A successful launch event will have been designed based on the required outputs and the available inputs (people, project status) and the input-process-output (IPO) tool is a good way to do this (Figure 3-11).

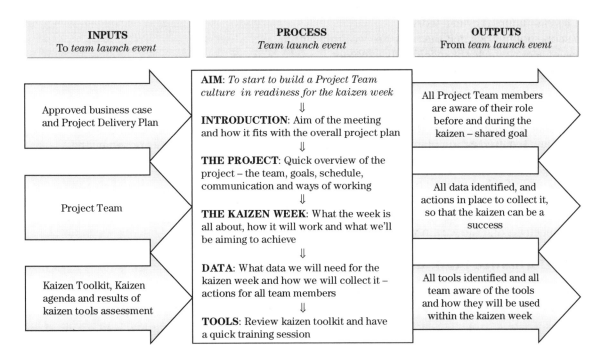

Figure 3-11 Example project launch IPO

The IPO (Figure 3-11) describes the launch event for a Project Team who are due to be using lean six sigma to improve their logistics business process. The Project Manager has decided to use a kaizen methodology (as used in the case study project at the end of this chapter, page 131). The kaizen is a full-time accelerated delivery methodology which uses a structured process and a set of lean six sigma tools to improve processes.

Performance

In order to have a performing team there needs to be a consistent understanding of the teams purpose as linked to the project vision of success. It is useful to have this vision very visible within the team working area and/or within team meetings. In this way it is a constant reminder to the team about what they are trying to achieve. In addition the team need to be aware of the ground rules for team operation, these should also be visible in team working areas:

- Acceptable behaviour in meetings or in open plan team work areas.
- Appropriate behaviour in external situations when they are representing their team.
- The values the team hold, such as delivering on commitments, treating all ideas fairly, listening to alternate view points.
- The way information is distributed; the method and frequency of communication.
- The level of confidentiality.

Each team member should have a clear role, bounded responsibilities and a set of performance targets which drive them to behave appropriately. Within any set of responsibilities limits also need to be set:

- The financial approval limits that a person has.
- The budget a team has for training.
- The decisions which they can or cannot take.
- The level of hours they can work.
- The scope boundaries that the team are working within.
- Areas of scope which are specifically excluded.
- The information they can access.

Summary

A Project Manager should review the effectiveness of the team start-up and then conclude by developing any required action plans. A Project Manager needs to reduce the risks posed by the team to as low as possible. Mitigating actions may include:

- Replacement of selected team members.
- Additional team building events to reinforce specific parts of the start-up plan.
- Additions to the team training plans, role profiles or team processes to support clear articulation of team objectives and the way in which they are to be achieved.

Team performance planning

During project planning the Project Manager needs to set up processes which will support the team in being high performing during the delivery of the project. There are three key areas to be considered:

- Culture
- The individual
- Performance management

Team culture

Culture, as discussed on page 50, is a combination of how we behave, the values we hold and the things we do. The development of team culture is no different. It is a function of how team members behave, the values that team members have and the activities they perform. Therefore in terms of planning team performance a Project Manager wants to generate and keep an appropriate environment, or culture, within which the team will perform.

A Project Manager would therefore plan to audit the environment by asking the team key questions during delivery which will demonstrate behaviours, values and actions. Typical questions which the Project Manager might ask are:

- **Behaviours:** Does the team behave in a way which matches the agreed ground rules? How do team members behave during conflict or disagreement? Do team members behave in a way which supports other team members?
- **Values:** Does the team have a clear and engaged vision of success? Is everyone treated with respect? Are team goals integrated into individual development plans?
- **Actions:** Do the project deliverables match the quality, quantity and functionality requirements of the project? Are decisions made considering individual, team and business goals?

The way in which the Project Manager behaves will have a strong impact on the project culture. If he is consistent and his actions and words match then it will generate a positive, trusting culture. If, on the other hand, the Project Manager says one thing and does another then a culture of mistrust can build up. This again links to the continuum of empowerment (page 227) and supports a Project Manager in managing performance in line with his articulated level of delegation and empowerment.

The individual

When delivering a project it is easy to forget that there are individuals involved. So much of the work completed at project start-up is focused on developing a shared understanding and a shared goal and way of working, that sometimes the individual is forgotten. However, team success needs individuals to feel and be successful.

An individual team member needs confidence in:

- The value in their role.
- Their value as an individual.
- Their value as a member of a team.

A Project Manager should have a plan for individual development and management throughout the project. This may mean using external mentors or line managers in the business who focus on individual development planning. The process should ensure that development actions align to current responsibilities.

Team performance management

A system to measure team performance should be set up at the start of a project and be capable of highlighting trends which will impact project success. Often measures systems are carried out at a number of levels linked to the scope definition (page 116):

- **Overall project goals**: A Project Manager would typically keep the team informed of the progress of CSFs and deliverables, using such tools as risk rating (page 157) and earned value 'S' curves (page 201).

➤ **Sub-project activities:** Within sub-teams or project work streams very specific metrics might be tracked for communication to a specific sub-set of the overall team. For example, a team looking at the procurement of items within an overall construction project might track how each team member is progressing a specific set of purchase orders.

➤ **Tasks:** Specific metrics might only look at individual tasks or individual categories of team activity. For example, in a construction project, team performance might be measured by the volume of concrete poured per day or metres of pipeline laid.

Apart from monitoring progress, activities or scope, team performance should be measuring:

➤ Number of team-related issues in period.
➤ Level of team morale.
➤ Effectiveness of relationships with key stakeholders as well as each other.
➤ Effectiveness of decision-making processes.
➤ Ability to adhere to team ground rules.
➤ Effectiveness of team leadership (at all levels in the project organization).

The goal of team metrics is to give the team sufficient data to understand how the team are progressing towards the shared goal and what they need to do to continue to support this.

Short case study

A product is to be launched and a complex design for the organization of the programme has been developed in order to pull together all the required internal resources. The programme is made up of six sub-projects and has a complex ownership of objectives, which is directly linked to the matrix management processes within the product development organization.

➤ Figure 3-12 shows this complexity – the programme is managed as a matrix with Product Project Teams (cross-functional) made up of R&D and/or manufacturing personnel with other functional responsibilities outside of the scope of this team.
➤ A programme of one core project and six sub-projects with a complex interdependency.
➤ All team members have one or more programme roles *plus* one or more BAU roles within functional areas.
➤ Each of the main role types within the organization chart has been developed (Table 3-11) as a precursor to formal launch when more detailed decision-making processes will be developed and agreed.

The design of the organization was able to accommodate the interaction of very senior functional executives with less senior Project Managers delivering a specific segment of a large and sensitive programme.

The fact that the more senior team members were reporting to a less senior colleague was never an issue because everyone realized that the organization was an appropriate way to organize the available and required skills and capabilities.

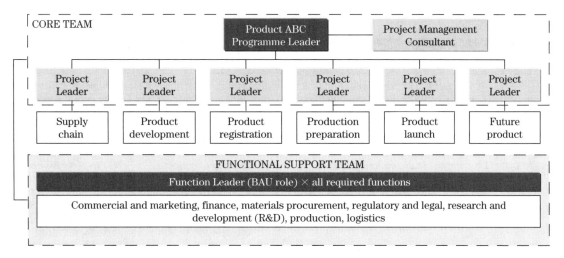

Figure 3-12 Case study programme organization

Table 3-11 Case study programme role profiles

Planning Toolkit – Project Team Role Profile			
Project: *Product Launch Programme*		**Sponsor:** *Name 1*	
Date: *Month 1*		**Project Manager:** *Name 2*	
Role	**Description**	**Performance metrics**	**Key relationships**
Product ABC Programme Leader	Responsible for the successful delivery of the programme via management of the core team Development of an appropriate programme delivery strategy that ensures risks are minimized and success potential is maximized Alignment of all Project Teams as defined within the PDP	➤ Realization of programme benefits ➤ Product ABC Core Team management – zero issues ➤ Project alignment – zero issues ➤ Sponsor satisfaction rating	➤ Programme sponsor ➤ Product ABC Supply Director ➤ Manufacturing Site Director ➤ Product ABC Development Director ➤ Product ABC Core Team ➤ Project Leaders
Product ABC Core Team	This team comprises all Project Leaders and all functional leads The team meets every two weeks to review project progress and the current risk profile They assign resources and manage any matrix conflicts They communicate with selected stakeholders	➤ Delivery of the programme objectives ➤ Programme risk rating drops no lower than amber ➤ Minimize project conflicts ➤ Stakeholder satisfaction rating	➤ Product ABC Programme Leader ➤ Functional Leaders ➤ Project Leaders

(continued)

Table 3-11 (Continued)

Planning Toolkit – Project Team Role Profile			
Project:	*Product Launch Programme*	**Sponsor:**	*Name 1*
Date:	*Month 1*	**Project Manager:**	*Name 2*
Role	**Description**	**Performance metrics**	**Key relationships**
Project Teams	Each Project Team requires a Project Leader, whose role is to ensure that functional scope/deliverables are defined and any cross-functional interdependencies are incorporated into PDPs	⟫ Delivery of project objectives ⟫ Project risk rating drops no lower than amber ⟫ Project deliverable progress meets plan	⟫ Product ABC Programme Leader ⟫ All other Project Leaders ⟫ Manufacturing Site Directors (as appropriate to the sub-project) ⟫ Other business leaders (as appropriate to the sub-project)
Functional support group	A set of individuals delivering specific areas of technical expertise to the Project Teams: ⟫ Manufacturing process ⟫ Product formulation ⟫ Supply chain logistics ⟫ Materials supplies ⟫ Regulatory/legal ⟫ Commercial and marketing ⟫ Long term demand modelling ⟫ Financial	⟫ Delivery of functional objectives ⟫ Functional risk rating drops no lower than amber ⟫ Project Leader satisfaction rating ⟫ Minimize functional resource issues across the sub-projects	⟫ Programme sponsor ⟫ Product ABC Programme Leader ⟫ Manufacturing Site Directors (as appropriate to the function) ⟫ Other business leaders (as appropriate to the function) ⟫ Product ABC Core Team ⟫ Project Leaders
Project Management Consultant	Provision of project and programme management expertise specifically linked to the development of the PDP Development of project management capability profiles across all sub-projects and assessing any risk associated with gaps identified Perform periodic Core Team and project health checks	⟫ Product ABC Programme Leader satisfaction rating ⟫ PDP delivered on time (programme)	⟫ Product ABC Programme Leader ⟫ Product ABC Core Team ⟫ Project Leaders

Project type

Although many project management tools and techniques are inherently generic and therefore useful for all types of project, there is still a benefit in considering the particular challenges of a specific type of project. However, all too often projects are categorized in a prescriptive manner and given a general label linked to their main focus.

For example, the upgrade of a laboratory in a research facility might be categorized as a small capital engineering project as it will be delivered by the engineering department. It then is managed within this category following a typical engineering project roadmap: design, procure, install, commission and handover. In some respects there is no issue here except for the fact that a major part of the upgrade could be linked to a new system of automation within the laboratory. Systems projects have a slightly different roadmap with different stage gates and time-scales: user requirements specification, functional requirements specification, hardware and software design, installation and factory and site acceptance testing. A Project Manager needs to consider the various categories and pull together a roadmap which adequately describes the journey which the project must follow. Examples of some different project categories are shown in Table 3-12.

Table 3-12 Project categories

Project	Project category						
	Capital funded	Engineering	Compliance	Systems	New product	Business change	Expense funded
New pharmaceutical facility	×	×	×	×	×		
Chemical facility upgrade	×	×	×	×		×	
Logistics process improvement			×	×		×	×
New product development		×	×	×	×		×
New computer system	×			×		×	

The reason for reviewing the project category is linked to the potential impact it may have on the development and delivery of an appropriate PDP:

➤ **The project process** – different types of project go through different journeys. If a project is a mixture of categories, then those various journeys need to be integrated.
➤ **Use of organizational resources (money, people, assets)** – Within an organization there will be existing business processes dealing with, for example, changes to assets or ways of working (WoW), use of people and funding requests. Each business process will have an expectation of decision points required and this also needs to be integrated into the project journey.
➤ **Control methodologies** – different types of projects have different deliverables and distinctly different activities and this may direct a Project Manager to the most appropriate control tools as well as the most appropriate types of resources.

Project roadmap definition

The definition of the overall project journey can be achieved through the use of a project roadmap. This is a flow chart which describes the main stages and stage gates that a specific project needs to go through in order to be successful. Apart from being customized for a specific project type it can also be customized to work within a specific organization and therefore interact with specific business processes.

A roadmap should guide the Project Team through a project and as such is seen as a steering mechanism. It is typically made up of two main activities: project stages and project stage gates (decision points). Figure 3-13 shows an example of a project roadmap which describes the stages and stage gates that a new pharmaceutical drug may go through to be made, launched and sold.

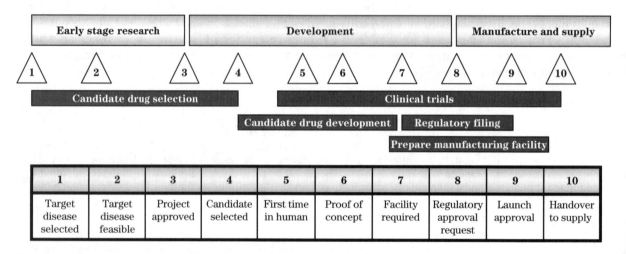

Figure 3-13 Example project roadmap – pharmaceutical drug development

Although this is a very high-level roadmap, it demonstrates that the organization has put in place key hold and/or decision points for drug development projects. This allows the company to choose to continue development, or not, depending on the results from key stages. In this way if a drug is not proving to be an effective solution to a target disease the development project can be stopped at an appropriate time.

Project stages

Figure 3-13 demonstrates that the first key aspect of a roadmap is the definition of the various project stages: what are the key steps the project must go through to deliver? These stages will be unique to the project type although most projects go through some forms of feasibility, design, implementation and handover.

Project stage gates

A project stage gate is a defined point at the start, during, or end of a project stage that requires some data in order to progress along the roadmap. It is usually clearer to have a stage gate at the end of a stage as a hold or decision point before progressing to the next stage. However not all stages require a stage gate.

A stage gate must have a purpose for either the project or the organization. In the example in Figure 3-13 the early stage gates were filtering out drug development projects which had little probability of successful launch. The earlier this is done the better for the organization as the resources are then available to work on other projects. A stage gate must have:

➤ A clear entry point – what should have been achieved before the hold or decision point.
➤ Acceptance criteria – appropriate information so that a decision can be made and/or a hold removed.
➤ A reason why it is needed – the value it adds to the project or organization.

Tool: Roadmap Decision Matrix

The key to the development of the right roadmap is the definition of the decision points in terms of 'what', 'when', 'how' and 'who'. The aim of the Roadmap Decision Matrix tool (Table 3-13) is to support Project Managers in defining stage gates to a level which is useable for themselves, the decision makers and the Project Team.

Table 3-13 Roadmap Decision Matrix explained

Planning Toolkit – Roadmap Decision Matrix				
Project: *<insert project title>*			**Sponsor:** *<insert name>*	
Date: *<insert date>*			**Project Manager:** *<insert name>*	
Stage gate	**Decision**	**Decision by**	**Decision when**	**Data needed**
<insert number>	*<insert decision to be made>*	*<insert name of decision maker>*	*<insert when the decision needs to be made>*	*<insert the data needed to make the decision in terms of meeting acceptance criteria or not>*

Stage gate

It is usual to number these in some way for easy reference.

Decision

There needs to be a clear reason for the stage gate and therefore a clear and defined decision to be made. Is the project waiting for an approval, for a go/no go, or for some external information before proceeding to the next stage? The decision should clearly indicate the next steps for all decision outcomes. For example, if a project has a go/no go stage gate then: 'go' might mean 'move to the next stage' and 'no go' might mean 'stop all project work now'.

Decision by

Be clear who can make this decision. It may be an individual in a specific role or it may be a team or group who have specific authority within the organization such as funding approval. The decision makers need to understand the decision they are making, the resources they are approving and the mechanism to do so. For example, a project go/no go decision might be related to its priority within a portfolio of projects rather than just being considered on its own merit. The decision to 'no go' may be because the organization does not consider the project a high enough priority even though the project makes business sense.

Decision when

There needs to be a clear definition of when in the roadmap the stage gate occurs and it is usual to link a stage gate to a specific project stage. These can then be used during schedule development and estimates of the stage gate timelines can be made.

Data needed

There should be a clear statement of what supporting data should be available in order for the decision to be made. In some cases this may be a specific deliverable which collates specific information or it may be a specific performance measure. The aim of the data is to prove that specific acceptance criteria have been achieved or not. One way to generate this level of definition is to develop an IPO for each stage and then to check that the outputs from one stage deliver both the required outputs for a decision and the required inputs for the next stage.

Finally, the Roadmap Decision Matrix can also be used as a part of the control strategy to track the achievement of the stage gates as the project is delivered.

The benefits of a stage gate approach

There are many reasons why adopting a stage gate approach is beneficial to both the organization and the project:

- As a project progresses from kick-off to completion there is an increase in the use of organizational resources. Stage gates are a method to check that the resources are being deployed appropriately for maximum organizational benefit. They are a check to see that the project is still able to meet organizational needs.
- Stage gates provide clear milestones for tracking at a project and organizational level. In addition, for similar projects in an organization they assist in generating benchmark data which can help forecasting and planning of future projects.
- Stage gates support schedule development and assist in defining the internal dependencies and thus definition of the project critical path (page 191). They are also useful in reviewing schedule risk and thus protection of the schedule through use of schedule buffers (page 195).

The project roadmap is a way to develop a customized project process which can consider the specific needs of a project or project type. Therefore the ultimate benefit of this approach is the way it builds in a 'lean' approach, only doing activities which add value to the project (page 194).

Short case study

A logistics organization wanted to encourage business improvement ideas but equally wanted to ensure that they optimized the use of resources to deliver the greatest benefit for the company. Figure 3-14 shows the project roadmap developed for change projects within this organization and Table 3-14 explains the stage gates.

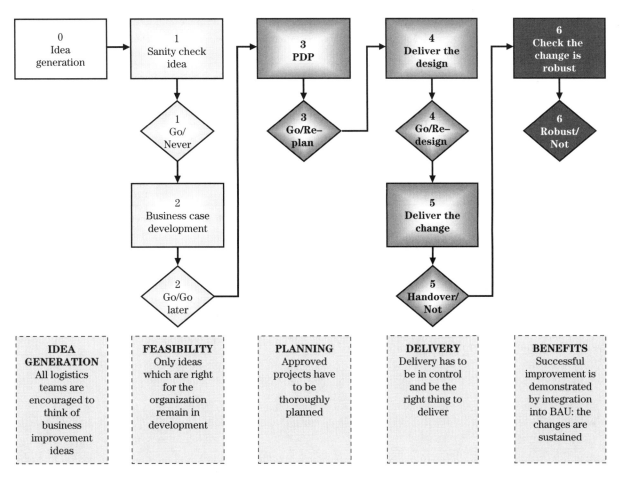

Figure 3-14 Example project roadmap – logistics change process

Table 3-14 Example Decision Matrix

Planning Toolkit – Roadmap Decision Matrix				
Project: Logistics Change Management			**Sponsor:** Fred Bloggs	
Date: Week 1			**Project Manager:** Martha Smith	
Stage gate	**Decision**	**Decision by**	**Decision when**	**Data needed**
1	Go or never	Management team	After feasibility work	Benefits of the idea Link to the business goals
2	Go or go later	Management team	After project defined	Business case
3	Go or re-plan	Project sponsor	After PDP complete	PDP
4	Go or re-design	Project sponsor	After design complete	Design (fully documented) Change delivery plan
5	Handover or not	Project sponsor	After change delivery	Deliverable status report Customer handover certificate
6	Robust or not	Project sponsor	After benefits delivery	Benefits tracking report Sustainability tracking report

The organization was able to shut down poor ideas quicker, before significant resources had been used, and to focus efforts on fewer, more robust business improvements that delivered significant benefits for the business. In addition the process better engaged the logistics operations teams, who were a fundamental part of the sustainability of change.

Funding strategy and finance management

Following approval of the business case there is inevitably a parallel approval of funds of some kind, either capital (CAPEX) or expense (OPEX). Within any organization there will be strict procedures which need to be followed to allocate and spend funds of any type. A Project Manager needs to work closely with 'accounts' in order to understand how to manage approved funds for the best outcome for the company. He would therefore develop a funding strategy. This will include legal considerations, taxation, management of the company balance sheet and company profit and loss account.

Finance strategy

Typical issues which need to be addressed by a finance strategy are:

- Funding approval.
- Types of funding.
- Financial governance.
- Finance management.
- Project impact.

Funding approval

Within some organizations business case approval may only approve sufficient funds to get to the next project stage gate. Within other organizations full funding may be approved right from the start. Different processes may also require a different level of project documentation in order to support approval. For example, within major capital engineering projects there is a phased approval of funds linked to the accuracy of the cost estimates presented (Figure 3-15).

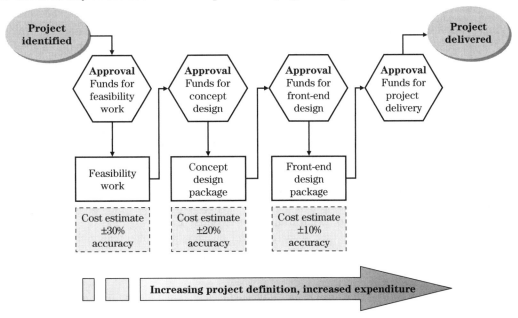

Figure 3-15 Engineering project funding approval process

The organization using the processes described by Figure 3-15 is maintaining central control of funds for as long as possible. Once it allocates full funding to a project then these funds become unavailable to the company and they want assurance that this is the right place to allocate the money.

Types of funding

Most projects are funded either entirely from capital or from operational budgets (expense). The choice of which is usually an organizational decision. However, accountancy principles are commonly applied and generally:

- Capital funding would be used to acquire new assets which are to become new organizational resources, which will then generate future benefits for the company.
- Operational funding (or expenses) would be used to maintain current assets.

Whether a project is capital or expense funded there are common elements to each:

Escalation funding

Often a business will set aside additional funds to compensate for any general market cost increases. For example, a major capital project requiring significant steelwork may need to consider when it intends to purchase the steel and if the cost estimate in the approved capital budget needs an escalation allowance. In any event most organizations do not consider escalation funds as an integral part of the project budget. This means that if the escalation does not occur then the project cannot use the funds for any other item in the project. For this reason, and for smaller areas of escalation, a Project Manager may choose to itemize this 'cost risk' under contingency.

Contingency funding

When funding is applied for it is usually based on a cost estimate of a specified level of cost accuracy. However, just because the cost estimate accuracy is ± 10%, for example, does not mean that a 10% contingency is either required or justified. Contingency funding is usually considered funding to cover areas of calculated risk. This is covered in more detail in the section on control planning (page 186). Contingency funds are usually considered an integral part of the project budget, but how they are approved for use is subject to differing rules depending on organizational processes.

Some projects use more complex funding arrangements as it links to their overall taxation or financial situation for any one year. For example, a company was going through a merger process and needed to relocate people from each company into one new head office building. The overall funding types used for each part of an overall programme were segmented into four main types (Table 3-15). The merger costs were considered exceptional costs that should not be included within annual performance reviews (by management and shareholders). The finance manager also considered separately which project costs were to be capitalized or expensed.

Table 3-15 Example – funding strategy

| | | Merger related | |
		Yes	**No**
Project related	**Yes**	➧ Asset disposals and write-offs associated with the main move to the new building ➧ New assets required as a result of the move (capitalized) ➧ Final move costs (moving assets)	➧ New office building (already has a capital budget) ➧ Major asset disposals already in progress
	No	➧ Relocation costs (people and IT infrastructure) ➧ Redundancy costs (for those unable to relocate)	➧ Any office moves linked to BAU activities

Financial governance

Due to the rules under which organizational finance needs to operate it would be usual to have some liaison with the finance department. The role of this support resource would be to ensure that the project finances are applied for, spent and generally managed in line with organizational and legal requirements.

For larger projects the part-time role may be replaced with a formal role in the Project Team: the project accountant. The typical types of activities which this role fulfils, as separate from general cost control, are:

➧ To advise on timelines for funding requests and ensure that cost plans are in the appropriate organizational format.
➧ To interpret financial rules and business processes.
➧ To ensure that all new assets are appropriately recorded in financial systems.

In some regards the financial governance is not about what you spend money on, rather it is about how you take account of that spend within the organization's accounting system.

➧ *Balance sheet* – the status of an organization's financial position: the assets and liabilities are stated.
➧ *Profit and loss account* – the performance of an organization over a specific time period: the revenue, expenses and profit are stated.

Finance management

Project finances should be tracked so as to support the business and this will mean:

➧ **Cash flow forecasts** – within a business the amount of finance used in any one accounting period is an important issue particularly when the project is over a number of years.
➧ **Project expenditure** – in line with cash flow forecasts, particularly at year end when annual accounts are being prepared.
➧ **Asset tracking** – both buying new assets and disposing of older ones and also forecasting when assets will be on line and available to generate revenue.
➧ **Accurate year end accruals** – knowing what funds are committed but have not yet left the company.
➧ **Contingency and escalation run-down** (page 189) – knowing what funds allocated to the project won't actually be needed.

Project impact analysis

Organization finance policies can impact project delivery strategy and needs to be reviewed. These policies can impose constraints such as:

- Inability to purchase any asset until full funding has been approved. For some long lead items this may directly impact the critical path and therefore extend the project schedule. In this scenario the organization may allow the purchase of the design of the asset and/or the purchase of the materials from which the asset can be made.
- Inability to purchase raw materials for the start-up of a new asset as they are to be procured from expense budgets not yet set-up or approved. In this scenario the start-up materials may need to be purchased from project funds and then expensed at a later stage.

For this reason a PDP should include definition of a finance strategy which aligns with the project delivery strategy.

Tool: Finance Strategy Checklist

Inexperienced Project Managers often ignore the implications of project financing as they focus on the traditional planning themes: cost and schedule development. The aim of the Finance Strategy Checklist (Table 3-16) is to highlight any financial issues which may impact the development of the PDP.

Table 3-16 Finance Strategy Checklist explained

Planning Toolkit – Finance Strategy Checklist	
Project: <insert project title>	**Sponsor:** <insert name>
Date: <insert date>	**Project Manager:** <insert name>
Funding types	
What types of funding are required for the delivery of this project? *<insert main type of funding: capital or expense or a mixture>* **Will either contingency or escalation funding be required?** *<identify the need for either contingency or escalation funds and explain why they are needed – for example, link to project type or level of cost estimation accuracy>*	
Funding approval	
In which part of the organization will the funding be approved? *<insert organization, department or name of approver>* **Is a funding approval process available?** *<outline the main activities which need to be performed in order to approve funds>* **Has any funding been approved?** *<insert funding level approved, by whom and against what supporting data>* **Is any funding still to be approved?** *<insert funding level yet to be approved, when approval will be requested and what data will be provided to support the approval>*	

(continued)

Table 3-16 (Continued)

Planning Toolkit – Finance Strategy Checklist			
Project:	*<insert project title>*	**Sponsor:**	*<insert name>*
Date:	*<insert date>*	**Project Manager:**	*<insert name>*
Finance management			
What are the liaison arrangements with the finance department? *<insert the name of the liaison finance manager and/or the assigned project accountant>* **Has a clear role been identified?** *<insert agreed role and/or RACI Chart>* **Have all financial procedures and specific issues been clarified?** *<attach key procedures such as reporting needs, asset management, cash flow and any specific constraints on the project linked to benefits realization or cash flow>*			
Summary			
Summarize the impact of the finance strategy on the project *<insert any constraints and the way that the project will have to deal with them>* **Will the project require any special funding arrangements in order to meet business goals?** *<insert specific project activities that may need special funding or finance management considerations>*			

Short case study

A project to build a facility to launch a new consumer product was due to be completed by year end, with production commencing shortly afterwards. There was an expectation that the revenue from product sales would have a significant positive impact on next year's profit and as a result the company share price has risen.

The Project Manager planned to complete the project on schedule and was working from a plan that he had developed based on the business case outlined to him. The plan stated that he was expecting to handover the facility at year end allowing the operations team to start 2 months of testing in the New Year, with production ramping up from 50% to 100% over the next 10 months. This would allow for a robust training strategy for the new operations team.

However, the financial business benefits of the project had never really been shared or challenged. As a result a finance benefit opportunity was missed. Had the Project Manager realized the finance assumption of having the facility 100% revenue producing by year end rather than complete by year end, he would have been able to fast track the build.

As it was, the capacity for year 1 was reduced to 65% of finance assumption with a corresponding reduction in profit for the company (and some loss in market share to a competitor). An after action review (AAR) highlighted that had the Project Manager been using a PDP that contained an outline of the high-level financial strategy he would have developed an entirely different project delivery strategy incorporating these needs, and thus avoiding any negative impact.

Scope definition

Once there is a clear business plan, a defined project roadmap and a start-up team in place, the scope development and detailed definition can begin. This is crucial data for the control plan and can generally be developed through the following steps:

- **Path of CSF development** – confirm the level 1 CSFs from which the project objectives have been defined.
- **Work breakdown structure (WBS) development** – develop a list of required project activities working down to lower-level CSFs which contribute to the level 1 CSFs.
- **Prioritization of objectives** – confirm the priority of each element of scope, described in terms of lower-level CSFs. Start to assign metrics and link to Project Team roles.
- **Quality, quantity, and functionality definition** – develop a clear understanding of interim and final deliverables.
- **Activity mapping** – develop a link between the benefits required and the associated scope needed to realize those benefits.
- **Scope risk assessment** – develop an understanding of how the quantity, quality and functionality could be at risk and develop action plans to mitigate these risks.

Throughout each of the scope definition steps, value definition and management is considered.

Value definition and management

Value is defined as 'a feature, condition, service or product that the customer considers desirable, and which is delivered to them when and where they want it' (Adams et al., 2004). A project should be delivering value to its customer and so the application of lean six sigma principles is appropriate, particularly when considering the scope definition. If a scope is 'lean' then each activity, decision and final deliverable adds value to the final deliverable: the tangible output which enables the customer-defined benefits to be realized. Anything else is waste (Figure 3-16).

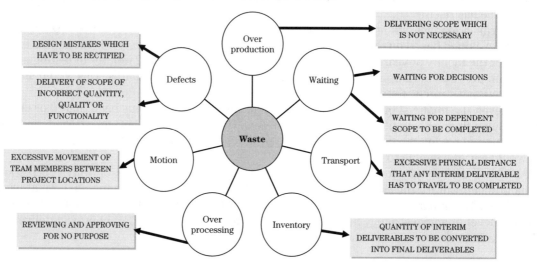

Figure 3-16 The seven types of waste applied to scope definition

The link between business benefits and project activities is analogous to the link between customer requirements (from a service or product) and the service or product specification:

➤ The scope defines what the project will deliver = service specification.
➤ Once delivered, the project should enable the benefits to be realized = customer requirements from a service.

A useful value management tool is a Kano analysis coupled with critical-to-quality (CTQ) trees.

Kano analysis

This is an excellent scope planning tool as it can be used to identify and classify customer requirements. The generic tool plots service specification versus customer satisfaction (Figure 7-9, page 272), but in a project context a Project Manager can plot scope versus benefits realization (which equates with business satisfaction). The scope can then be categorized into:

➤ **Must have** – without this scope the basic benefits metrics will not be realized and the business will not be satisfied.
➤ **More is better** – if we have more of this scope we will realize more benefits and the business satisfaction will increase.
➤ **Delighters** – if we deliver this scope the benefits realization will exceed expectations and the business will be delighted.

For example, if we look at a process improvement project the scope brainstorm conducted by the team can be analysed and plotted on a scope kano analysis chart. The final scope is then defined as all 'basic' scope plus any 'more is better' scope which delivers the required level of business benefits (Figure 3-17). All scope above the benefits target level is excluded as not required once it is checked to have no dependency to the required scope.

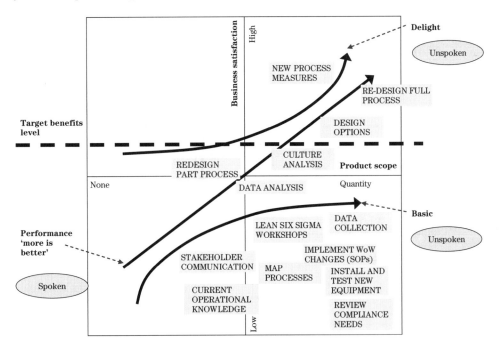

Figure 3-17 Example Kano Analysis as applied to project scope definition

The project in Figure 3-17 is very simple but it identified that post implementation the new process measures should be developed by the BAU team and not the project. Another key feature of this type of analysis is that it highlights the parts of the project scope that the customer is not aware of and would never ask for.

For example:

➡ A customer has identified that the scope should include four new stainless steel vessels.
➡ The Project Team know that the scope must include vessel specification and also supplier selection if the required benefits are to be realized.

CTQ Trees

A CTQ is defined as a key feature by which a customer evaluates the quality of a product or service the feature is 'critical to quality'. Again this is analogous to scope features which enable the business to assess whether the project is likely to deliver the required benefits.

In many respects a CTQ Tree is similar to the collation of a benefits map and an activity map. It links critical scope to critical benefits, features which the business uses to assess project success. A simple example of a CTQ Tree is shown in Figure 3-18 for an accountancy system improvement project. The analysis helped to widen the scope from purely IT system development into a business management process improvement (development of a kanban and an SOP).

In terms of value management, once a scope has been defined based on value principles then this must be delivered. The control of value delivery is a key part of the project control plan (Chapter 4).

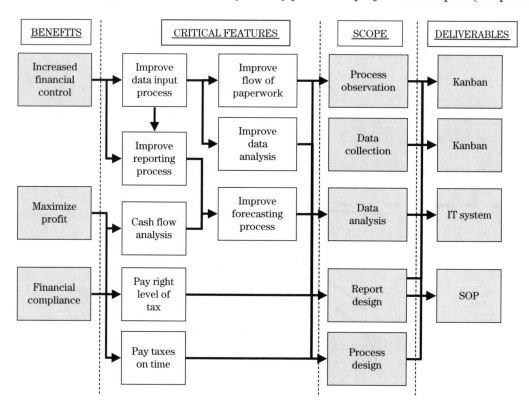

Figure 3-18 Example CTQ Tree as applied to project scope definition

Path of CSF development

A key step in identifying categories of scope is to develop a path of CSFs. A CSF is an identifiable action/activity that is quantifiable/measurable. It is 'critical' because without it the project vision of success cannot be achieved. CSFs do not always have finish-to-start dependencies as implied by the linking of CSFs into a path; they can be completed in parallel across the life of a project in some form or another.

A set of overall project CSFs (level 1) is referred to as the project 'Critical Path of Success' (Melton, 2007). They are usually generated through the following typical steps:

Step 1: Vision of success

Generate a clear vision of success for the project. This will usually have been completed in earlier planning phases (business planning) and link in to the approved business case for the project.

Step 2: Brainstorming

Use the team to generate ideas on what would help achieve the vision of success. Follow the 'rules' of brainstorming and don't allow any selection or judgement of the ideas at this stage.

Step 3: Categorization

Allow the team to sort the ideas into similar groupings – these are potential CSF categories. In effect at this stage an Affinity Diagram has been developed (An Affinity Diagram is a visual technique for organizing and grouping information.). It is expected that there would be about 5–10 categories at this stage.

Step 4: Category descriptions

Consider carefully what each category is about – give it a name and write a short description.

Step 5: Interrelationship analysis

The team now need to systematically identify any cause–effect relationships between the categories. This is usually done visually by writing all potential CSFs in a circle on a large flip chart/white board. Ask the team whether CSF 1 causes or influences CSF 2, then CSF 3, until CSF n, where n represents the number of potential CSFs. Then move on to CSF 2: does this cause or influence CSF 3, CSF 4, CSF n. Lines should be drawn connecting potential CSFs with arrows showing the direction of influence. Be sure to test these relationships: 'does this CSF really influence this CSF?'

Step 6: Drivers and outcomes

For all potential CSFs count up the number of 'in' and 'out' arrows.

- The item with the most 'out' arrows is the *key driver* for the project.
- The item with the most 'in' arrows is the *key outcome* for the project.
- Key drivers and outcomes can be defined **if** their score is significant. They would need an 'out' or 'in' score of more than half the total number of categories. A score of a third or above is only significant as a secondary driver or outcome **if** it is either **only** a driver or an outcome.

The analysis will give a clear indication of whether a potential CSF is a driver or an outcome. It can also demonstrate that a potential CSF can be of no significance to the project and is therefore not critical to success.

Step 7: Path of success definition

The Path of CSFs can be developed from this inter-relationship analysis as most potential CSFs can be divided into one of the two types: drivers or outcomes:

➤ Drivers become CSFs (either on their own or in combination with others).
➤ Outcomes become measures of achievement of the vision (and may cause the vision statement to be modified or become more detailed).

At the end of the analysis there is a set of SMART CSFs and a vision which can be measured (SMART = specific, measurable, achievable, relevant and time based).

For example, the case study at the end of this chapter (page 131) contains an example of a path of CSFs as linked to a project vision (Figure 3-25, page 139).

➤ This path of success was developed using the above process: 11 potential CSFs were identified and through both categorization and combination the actual number of defined level 1 CSFs was reduced to six. The vision was modified to include the key outcomes, making it SMART.
➤ The level 1 CSFs were used to develop individual activities that needed to be done (Table 3-24, page 140).
➤ Dependency links were considered at this activity level: 'I need activity x to be complete before I can start activity y'.
➤ Lower level activities are referred to as level 2 or 3 CSFs.
➤ All scope was linked to a CSF ensuring that value was built in and managed.

A clear set of overall project CSFs is a valuable start point for the development of a work breakdown structure (WBS), which is the base input for the development of a project schedule.

WBS development

Developing a WBS is a typical methodology to segment the scope into bounded activities and deliverables. A WBS allows a Project Manager to segment scope types and this can mean that the same type of activity (level 2 or 3 CSF) may be completed in different parts of the WBS.

For example, an engineering project with a CSF to 'design in compliance to all appropriate regulations', would apply to a vessel design and to a drier design. Level 1 CSFs tend to be applied across a project and a WBS through the vertical details of a project (Figure 3-19).

In Figure 3-19 the design scope has been divided into three types of interim deliverable or activity. These in turn consider the quantity required to fully deliver the final deliverable: in this case a new manufacturing facility. Once the WBS is appropriately detailed it can be used within the project control plan to:

➤ Identify required resources – quality and quantity.
➤ Identify dependencies between activities and deliverables.
➤ Confirm alignment to the level 1 CSFs.
➤ Define quality, quantity and functionality requirements.

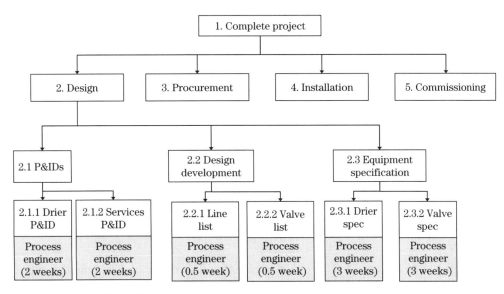

Figure 3-19 Example work breakdown structure

WBS is often initially shown as a tree diagram (Figure 3-19) and then converted into a list of activities which can be input into schedule, cost and resource plans. The goal of a WBS is to break a project scope into small enough pieces to enable robust delivery. The WBS in effect provides a framework from which the project work can be organized and managed.

A WBS can be deliverable oriented (Figure 3-19) or process oriented. In the case of the latter the WBS typically matches the breakdown of level 1 CSFs as would be shown within a Table of CSFs.

Quality, quantity and functionality definition

Once the CSFs are understood and the WBS defined, a review of scope quality, quantity and functionality should be performed.

Quality

Quality, in this context, is defined as the characteristics of an activity or deliverable which are required to meet a specified need. For example, a vessel specification needs to have very specific information in order to meet engineering standards; an SOP for a logistics operation needs to contain specific instructions so that materials move to the required location. All activities and deliverables within a project scope need be a specified quality and the scope definition process should address this.

Quantity

Quantity, in this context, is defined as the amount of an activity or deliverable which is required for a specified need. For example, within an engineering project it could be the number of each type of deliverable such as 4 vessels or 5 SOPs or 20 tonnes of concrete.

Functionality

Functionality, in this context, is defined as the performance criteria for an activity or deliverable. For example, within a product development project it could be the performance specification of the product: how it will perform or function for the user.

Scope categories

For each item of defined scope there should be a definition of its quality, quantity and functionality. In addition each scope item should be categorized:

- **Activity** – which will generate an interim or final deliverable.
- **Interim deliverable** – required to produce a final deliverable.
- **Final deliverable** – required for delivery to the customer to enable benefits to be realized.

Scope planning

Each item of scope has to be managed (in terms of planning delivery of quality, quantity and functionality) and there are various tools available to do this. An example is a quality matrix (Table 3-17). This type of tool focuses on interim deliverable management to support delivery of the appropriate quantity, quality and functionality.

Table 3-17 Example quality matrix

Deliverable	Deliverable quality management						Status	
	PE	ME	MD	IE	PM	Client	Revision	Date
Reactor module								
Reactor specification	M A1 F	D A1 C F	R C F	R C F	R A2	R A1 F	03	January year 1
Reactor P&ID	M A1 F	D A1 C F	D C F	D A1 C F	R A2	R A1 F	02	January year 1
Line and valve list	R F	D A1 F	M	R F	A2	F	02	January year 1
SOP	M A1	F	F	R F	A2	R A1 F	01	January year 1

Key to Table 3-17

PE = Process Engineer	M = holds master copy	D = technical development
ME = Mechanical Engineer	A1 = technical approval	C = needs access to master copy
MD = Mechanical Designer	A2 = commercial approval	F = needs final approved copy
IE = Instrument Engineer	R = technical review	

Activity mapping

Another methodology to support scope definition is activity mapping. This is a brainstorming process to develop the scope breakdown and is less structured than the WBS methodology. An advantage of using this approach is that a number of issues can be dealt with as scope is suggested:

- Alignment to a level 1 CSFs.
- Dependencies to other activities.
- Quality, quantity and functionality issues.
- Resource needs.
- Links to deliverables and from them to benefits.

An activity map is often used when the scope is more easily aligned to CSFs than work types as in WBS. The output from an activity map can be presented as a WBS, an activity plan (Table 3-18) or as a network diagram.

Tool: Activity Plan

The aim of this tool (Table 3-18) is to support the scope definition and to ensure robust alignment of scope to the project CSFs. The data contained within the activity plan is an input to the project control plan as it supports identification of:

- Dependencies between activities.
- Volume of scope.
- Required needs in terms of resource capability and capacity.

Table 3-18 Activity Plan explained

Planning Toolkit – Activity Plan			
Project: <insert project title>		**Project Manager:** <insert name>	
Date: <insert date>		**Page:** 1 of 1	
CSF	**Activity**	**Quality, quantity, functionality**	**Dependencies**
<insert CSF number>	<insert activity from work breakdown structure or activity map>	<insert definition of the activity in terms how much, to what specification and to what performance criteria>	<insert any other activity which must be complete or at some stage of progress in order for this activity to commence>
	<insert activity a>		
<insert CSF x>	<insert activity b>		
	<insert activity c>		

It would usually be supplemented with a project path of success. An example is contained in the case study at the end of this chapter (page 140).

CSF

Identify the level 1 CSF which links to the activity. Sometimes an activity has an alignment to more than one CSF and this should be noted. The latter is more likely to occur if a WBS methodology has been used to generate the activity listing rather than an activity map.

Activity

This is an identifiable, bounded area of scope. The activity may be one of three types:

➤ Activity leading to another activity – such as *data collection* that links to *data analysis*, which links to *solution development*.
➤ Activity producing an interim deliverable – such as *data collection planning*, which produces a *data collection plan*.
➤ Activity producing a final deliverable – such as *write new SOP*, which leads to the development of *an SOP* that is handed to the user at project completion.

Throughout the use of this tool it is important that only scope which adds value is included. For example, any interim deliverable which has no link to the final deliverable is waste.

Quality, quantity, functionality

For each area of scope there should be a clear definition of its quality, quantity and functionality requirements.

Dependencies

Activities link together like a path from the start to the end of the project. It is important to understand the links, define them and then manage them.

Prioritization of objectives

A scope developed in a lean way will contain only those activities and deliverables which add value. Therefore 100% of the scope is necessary for project success. However, in order to support decision making during the delivery of a project, (when change occurs or when risks become live issues) there needs to be an understanding of priority. This is usually defined in the control plan (page 145); however, it is pertinent to briefly discuss scope prioritization here. The short case study, below, demonstrates how scope can be prioritized through prioritization of high-level objectives.

Short case study

A project to launch a new product has identified four CS Fs and from these a series of activities and objectives have been defined:

- *CSF 1*: Facility available to produce required capacity.
- *CSF 2*: Supply chain can meet launch needs.
- *CSF 3*: Markets are ready.
- *CSF 4*: Product output meets quality requirements.

As all four CSFs are critical by definition the Project Manager has confirmed that the overall priority of objectives should be:

- *Schedule* – highest priority for maximum impact of the product launch on the market.
- *Quality* – next highest priority in order to get sustainable market penetration and this will be supported by high-quality products.
- *Capacity* – a medium priority as long as enough launch stock is available.

This would ensure that scope which was required for the launch had a higher priority than scope required for the sustainable supply post-launch. Clearly both were required for overall project success, however, the Project Manager needed a decision-making framework 'just in case'. A Table of CSFs (Table 3-19) was then developed with all level 3 CSFs prioritized against these criteria. The aim of the level 3 CSFs is to generate metrics which can be tracked in order to increase confidence that the critical milestone will be achieved:

- *Critical milestone* – date when the high-level objective is to be achieved.
- *Key milestone* – date when an activity crucial to the completion of the critical milestone is to be achieved.
- *Objective tracking metric* – a feature of the high-level objectives (level 2 CSFs) which can be measured as a trend throughout the life of the project (e.g. within monthly progress reports).

Table 3-19 was fundamental in defining all aspects of the scope and in agreeing responsibilities and priorities.

Table 3-19 Short case study: Table of CSFs

Project Management Toolkit – Table of CSFs				
Project: Product Launch			**Date:** Month 1	
Critical path of success				
To successfully launch product XYZ into two new markets from a new manufacturing facility in time for the next peak in seasonal demand				
CSF definition				
Scope area (CSF level 1)	**Objective tracking metric (CSF level 2)**	**Critical milestone (CSF level 3)**	**Accountable for CSF level 3 delivery**	**Priority (within scope area)**
CSF 1 Facility available to produce required capacity	Facility modified and ready for production	Facility modification complete (DATE)	Facility Project Manager	HIGHEST
		Facility tests complete (DATE)		
		Facility handover to production (DATE)		
	Facility capacity trends	Capacity per line meets design (DATE)	Facility Commissioning Manager	MEDIUM
		Rejects rate meets design (DATE)		
CSF 2 Supply chain can meet launch needs	Materials stock to manufacturing facility	Completion of launch stock (DATE)	Logistics Director	HIGHEST
	Warehousing capacity	Warehouse launch capacity ready (DATE)		MEDIUM
	Customer order processes	Process goes live (DATE)		MEDIUM
		Customer orders for launch (DATE)		
	Launch stock dispatched to markets	Completion of launch stock (DATE)		HIGH

(continued)

Table 3-19 (Continued)

Project Management Toolkit – Table of CSFs				
Project: Product Launch		**Date:** Month 1		
To successfully launch product XYZ into two new markets from a new manufacturing facility in time for the next peak in seasonal demand				
CSF definition				
Scope area (CSF level 1)	**Objective tracking metric (CSF level 2)**	**Critical milestone (CSF level 3)**	**Accountable for CSF level 3 delivery**	**Priority (within scope area)**
CSF 3 Markets are ready	Market demand trends	Agreed sustainable supply quantities vs accurate and regular forecasts (DATE)	Marketing Director	HIGH
		Agreed launch quantities vs accurate and regular forecasts (DATE)		
	Market regulatory approvals	Market 1 approval (DATE)	Marketing Director	HIGHEST
		Market 2 approval (DATE)		
CSF 4 Product output meets quality and demand requirements	Launch stock manufacture	Completion of launch stock build (DATE)	Production Director	HIGHEST
		Launch stock passed quality checks (DATE)		
	Sustainable manufacture	Capacity meets demand (DATE)	Production Director	MEDIUM
		Stock build meets demand (DATE)		
		Stock out per month		
		Reject rate		
		Line capacity		

Within this project there were problems with the product manufacturing process and the Project Manager had to develop a contingency plan which purely focused on meeting launch needs (recognizing that the development of the final sustainable supply would be delayed). Following launch on time there were a few stock outs in one or two markets but this did not impact overall market penetration. The project was successful but only because it focused on the high-priority objectives.

Scope risk assessment

During the development of the project control plan (page 157) risk assessments are conducted. One of these is a review of the risks to delivery of the scope and the various impacts this can have on delivering the project goals. Usually the main scope areas are reviewed against the highest priority goals (usually linked to cost, schedule and/or quality, quantity and functionality) and any risks identified at an early stage.

For example, a small engineering building project to install a new reactor into an existing facility has identified a number of risks associated with the three main scope items (Table 3-20). The risks are different depending on the progress of each scope item:

➤ *Stage 1*: Design scope.
➤ *Stage 2*: Construction and installation scope.

Table 3-20 Scope risk review example

Scope	Cost risks		Schedule risks		Technical risks	
	Stage 1	**Stage 2**	**Stage 1**	**Stage 2**	**Stage 1**	**Stage 2**
Reactor module	LOW Uses a standard design	LOW Confirmed quotation and vendor details on how to install	LOW Uses a standard design	HIGH Vendor has a reputation for late deliveries	LOW Uses a standard design	LOW Standard installation and standard use in process
Control system	HIGH Customized design	HIGH Likely on-site changes needed	HIGH Lots of design iteration needed	HIGH Typically late and high dependency on other activities	MEDIUM System functionality is checked at the factory	HIGH System functionality is checked before handover
Civil works	LOW Standard design for fixed fee	MEDIUM Some day works likely due to site conditions	MEDIUM Needs site information	LOW Using a reliable site contractor	LOW Standard design	LOW Standard works

The risk analysis allowed the Project Manager to assess cost, schedule and technical contingency plans so that he had a high degree of assurance that the scope would be delivered (to the appropriate quality, quantity, functionality requirements) whilst also meeting cost and schedule needs.

For any aspect of scope it is useful to consider 'what if' this scope item was below quality, less than 100% delivered or unable to perform as required. In this way appropriate mitigation and contingency plans can be developed.

Tool: Scope Definition Checklist

The aim of this tool (Table 3-21) is to ensure that a Project Team builds an effective scope. This tool checks that all scope definition principles have been covered and that there is a robust definition of scope in readiness for delivery. As a project control plan relies on the scope definition, it is sensible to do a final check during completion of the set-up plan, but before significant work is started on the control plan.

Table 3-21 Scope Definition Checklist explained

Planning Toolkit – Scope Definition Checklist			
Project:	<insert project title>	**Project Manager:**	<insert name>
Date:	<insert date>	**Page:**	1 of 1
Scope definition			
How has the scope been defined in this project? <insert description of the process which has been followed> **Is there a clear set of level 1 CSFs linked to an agreed project vision?** <insert yes/no and attach path of CSFs> **Is there a complete list of project activities, interim and final deliverables?** <insert yes/no and attach> **Do all activities and deliverables have a clear value for this project?** <insert yes/no and any actions to rectify as appropriate>			
Deliverable management			
Are the boundaries between packages of work (activities/deliverables) clear? <insert yes/no and any actions to rectify as appropriate> **Is it clear how the quality, quantity and functionality of activities or deliverables is to be managed?** <insert yes/no and attach document to demonstrate the process> **Is it clear how the activities or deliverables link to the project roadmap?** <insert yes/no and confirm any stage gate links>			
Scope risks			
Are there any quality risks? <insert yes/no and attach risk assessment and mitigation actions> **Are there any quantity risks?** <insert yes/no and attach risk assessment and mitigation actions> **Are there any functionality risks?** <insert yes/no and attach risk assessment and mitigation actions>			
Summary			
Summarize the status of the project scope definition <insert any issues and the way that the project will have to deal with them> **Is the scope suitably defined for the start of project control planning?** <insert yes/no and any further actions required>			

Scope definition

These set of checks test whether the scope is completed, that is adds value and that it has been generated appropriately delivering the required outputs: defined work packages. These are necessary for input to the control plan where the scope is costed, scheduled and resourced.

Deliverable management

In order to maintain the value defined during early scope definition it is important to check the boundaries between work packages to avoid duplication or potential gaps. An activity map is a good method to check this.

Scope risks

Specific risks related to the scope and their impact on the project need to be defined along with any mitigation plans. Some mitigation plans may impact cost or schedule.

Summary

It is important to conclude on the status of the scope definition as a key part of the completion of the set-up plan and an input to the control plan. The scope must be robustly defined so that it can be delivered and controlled during delivery.

Set-up plan case study – purchasing business process improvement

To illustrate the key points from this chapter an extract from a 'real life' PDP for a business process improvement project is described.

Situation

A recent review of an organization's ability to purchase key services or equipment has been completed and the metrics demonstrate that significant improvement is necessary.

- It can take extended periods to select a supplier and then even longer to get a purchasing contract agreed.
- The purchasing team has become larger but this is having little impact on the backlog of purchase contracts yet to be completed.
- The number of contracts which have expired is increasing.

A business case was developed and proved that without business process improvement the organization would be unable to demonstrate appropriate control of its external spend (a budget of approximately $100 million/year). A Project Manager was assigned and asked to deliver the business case as summarized by the Benefits Hierarchy (Figure 3-20).

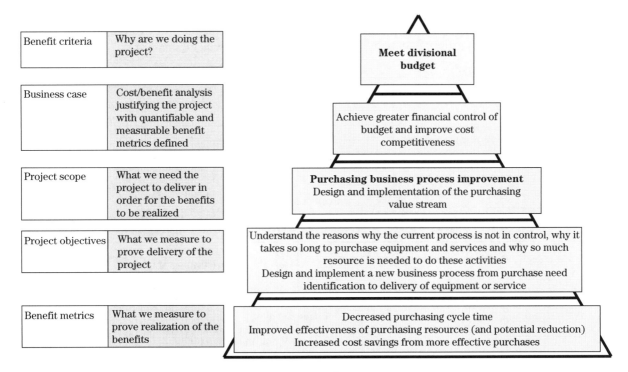

Figure 3-20 Benefits Hierarchy – purchasing business process improvement

The benefits required the design and implementation of a new business process which would add additional value to the business in three distinct ways:

- Get better at doing 'deals'.
- Do the 'deals' faster.
- Spend less effort in doing 'deals'.

It was therefore seen as a challenging project requiring a new approach within a fast-track time-scale: benefits needed to be delivered within 6 months to support achievement the end of year budget targets.

Set-up plan

The Project Manager had not completed a project within a purchasing environment previously; however, she felt that her background in generic business change projects was sufficient to be able to cope with the challenge. In addition, she was trained in the use of lean six sigma techniques and, based on the data presented in the business case, she felt that the use of such techniques would be crucial in quickly identifying and implementing the appropriate business process changes. Following assignment to the project she quickly formed a team and started to plan the delivery of the project. She used the development of a set-up plan as a structure for effective project launch.

Project organization

The Project Manager quickly assessed that the resources currently allocated from the business (herself plus one change manager and one purchasing representative) was not adequate. Initial planning for the project organization was conducted in liaison with senior stakeholders, and through negotiation additional resources of appropriate capability were eventually released (Figure 3-21).

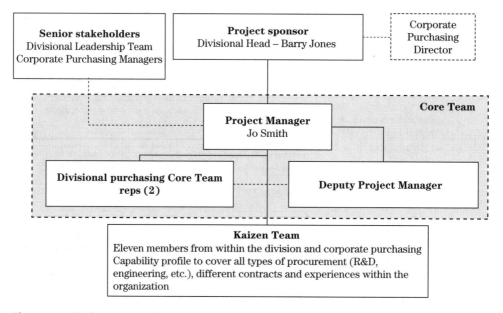

Figure 3-21 Project organization – purchasing business process improvement

The project organization (Figure 3-21) was designed to minimize use of resources so that operation of BAU was minimally impacted by the early stages of the project. It comprised four main parts:

- **Sponsor** – the person in the organization responsible for the $100million budget.
- **Senior stakeholders** – the people who have responsibility for operation of the existing purchasing business process and/or who manage significant elements of the $100million budget.
- **Core team** – a small group of people made up of 50% project/change management capability and 50% purchasing business process capability who will complete the majority of pre-delivery work.
- **Kaizen team** – a larger group of people from the purchasing business process who will be involved in the design and implementation of the new business process. They will in effect be designing changes to their own working environment.

The selection of team members was completed against a set of criteria as shown by the Team Selection Matrix (Table 3-22). This table shows the assessment of all 20 staff currently working within the business process.

Table 3-22 Team Selection Matrix – purchasing business process improvement

Planning Toolkit – Team Selection Matrix							
Project:	Purchasing business process improvement			**Sponsor:**		Barry Jones	
Date:	Week 1			**Project Manager:**		Jo Smith	
Potential team member	**Selection (yes/no)**	**Selection criteria**					
		Business change project experience	**Purchasing experience**	**Area of the business**	**Exposure to lean six sigma**	**Readiness for change**	**Available for kaizen week**
Jane	**Yes core**	Yes	10 years	All	Some	**High**	Yes
Claire	**Yes**	Minor	6 years	R&D	None	**High**	Yes
Tim	No	None	15 years	R&D	None	Medium	No
Robert	**Yes core**	Yes	5 years	Engineering	Some	**High**	Yes
Russ	No	No	5 years	Utilities	None	Low	No
Anthony	**Yes**	Minor	10 years	Utilities	None	Low	Yes
John	No	Minor	8 years	Engineering	None	None	Yes
Bruce	**Yes**	Yes	5 years	All	Some	Medium	Yes
Sarah	No	Minor	6 months	FM	**Yes**	**High**	No

(continued)

Table 3-22 (Continued)

Planning Toolkit – Team Selection Matrix							
Project: Purchasing business process improvement				**Sponsor:** Barry Jones			
Date: Week 1				**Project Manager:** Jo Smith			
Potential team member	**Selection (yes/no)**	**Selection criteria**					
		Business change project experience	**Purchasing experience**	**Area of the business**	**Exposure to lean six sigma**	**Readiness for change**	**Available for kaizen week**
Anne	No	Minor	4 years	All	None	Low	Yes
Jane B	**Yes**	No	3 years	Capital	Yes	**High**	Yes
Simon	No	No	6 years	R&D	None	Medium	Yes
Jeff	**Yes**	Minor	1 year	FM	None	**High**	Yes
Karen	**Yes**	No	6 years	FM	None	Medium	Yes
Caroline	**Yes**	Minor	2 years	R&D	Some	**High**	Yes
Henry	No	No	15 years	Operations	None	Low	Yes
Robert B	**Yes**	Minor	14 years	Capital	Some	Medium	Yes
Nathan	**Yes**	Yes	5 years	Operations	Some	**High**	Yes
Sue	**Yes**	No	18 years	All	None	Medium	Yes
Jeremy	**Yes**	Yes	2 years	Engineering	None	**High**	Yes
Summary							

In considering the 20 purchasing business process team members:

➥ Three were not available for the selected dates for project delivery via the accelerated kaizen methodology
➥ One was considered to be actively resistant and too much of a risk to be involved in the project
➥ Three were known to be passively resistant but their purchasing knowledge and experience would be invaluable and the kaizen would be an ideal mechanism to engage these individuals
➥ A larger than anticipated capability gap in terms of business change tools and processes was identified

Of the 20 individuals reviewed the Project Manager was able to select 11 main team members and 2 core team members. Based on a review of the scope during project kick-off, the roles for these two teams were developed further. This helped to manage team members' expectations and also ensure that everyone was clear on what was expected of them pre-delivery during BAU, delivery and post-delivery.

Core team role

The aim of the core team was to generate a strong project management culture aimed at delivering on time and within agreed benefits expectations. In addition, the core team needed 'insiders' from the business: those people who would actually be impacted by the changes themselves. Choosing two purchasing managers with high credibility supported 'buy-in' from their team members (both within and external to the Project Team itself) and also set in place a mechanism for future sustainability of the changes. This team had a detailed RACI Chart to ensure that the project management activities were all assigned as well as other activities.

All members of the core team had responsibilities leading up to the actual delivery phase, although they also had to continue with their 'day job'.

Kaizen team role

In order to improve a process, data is needed. The people best placed to collect and analyse this data are the people who will ultimately have to work within the new process. Therefore the team role was simply:

➤ Collect data in line with the data plan.
➤ Attend the kaizen.
➤ Complete any post-kaizen actions.

This meant that generally team members did not have major activities to complete in the 6-week planning phase which supported their release for the project.

In reviewing the background and experience of the selected team members it was clear that there was a capability gap related to the process and tools associated with a kaizen event. To understand the size of the gap a kaizen tools assessment was completed (Figure 3-22).

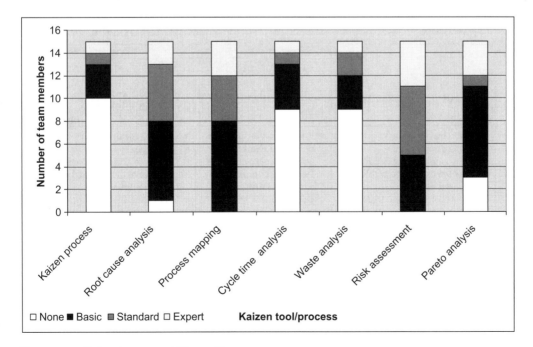

Figure 3-22 Kaizen team capability profile

The Project Manager knew from previous experiences that a successful kaizen event required a team with a high proportion of standard capability in these tools/processes. So, as a part of the team building process a training event was held with post training mentoring to reinforce learning.

Project roadmap

The Project Manager had some experience of business change projects and knew that the development of a roadmap was crucial in setting the overall approach from which to develop a control plan. Therefore a lean six sigma approach was proposed based around the DMAIC Process:

D = define – look at current business process, consider customer views.
M = measure – assess current performance.
A = analyse – find root causes of performance gaps.
 I = improve – identify solutions.
C = control – track and control the changes.

This is a systematic methodology requiring an experienced facilitator who guides a team of individuals with knowledge and experience of the business process being reviewed. However, it is only a framework and the Project Manager still needs to choose the most appropriate tools to identify and eliminate waste, and to identify and reduce variation.

It is difficult for organizations to make significant changes to their business processes if they are not able to allocate sufficient resources (usually the most limited commodity within today's organizations). Often change projects are resourced with part-time roles and significant external contract or consultancy support. This can lead to slow, non-sustainable and often unsuccessful change. The Project Manager was aware of all these, having had resource problems on much less complex change projects. She therefore decided to use a 'kaizen' methodology as one proven way to avoid these problems (Figure 3-23).

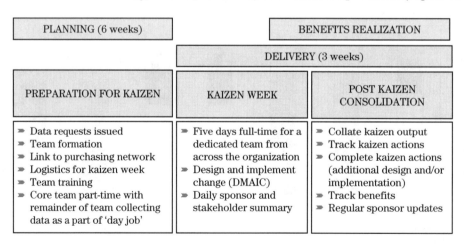

Figure 3-23 Kaizen methodology

- Kaizen (Japanese) – 'to make better' continuous, incremental improvement of an activity to create more value with less waste.
- Kaizen week – an accelerated, resource intensive but short time-scale methodology which requires stakeholder and sponsor involvement (decision making) at the end of each day in a week-long programme.

A kaizen methodology in this context means:

- Going through the DMAIC process in 1 week using a standard accelerated methodology on a bounded value stream.
- Assigning resources full-time for a short fixed period.
- Gaining senior stakeholder support for the outcomes of the kaizen, i.e. the improvements designed *will* be implemented.
- Using planning time pre-kaizen to collect sufficient data to ensure that the design is robust for current and future needs.

Figure 3-24 shows a typical kaizen timetable. Prior to the start data is collected and some pre-analysis is completed. During the week the DMAIC structure is applied and new processes are designed ready for implementation both during and following completion of the kaizen. There are typically two outcomes from a Kaizen:

- An approved, designed and implemented process change.
- Identification of an area which requires additional design and/or testing prior to implementation.

	Monday	Tuesday	Wednesday	Thursday	Friday
AIM	Data review and 'as is' analysis	Root cause analysis	Solution development	Solution development	Kaizen conclusions
TOOLS	» Data mapping » Activity mapping » Process mapping » Benefits mapping » Vision and path of success	» 5 Whys » Multiple cause » Fishbone » Time-Value maps » What if	» Process mapping » Benefits specification	» Benefit matrix » Simulation analysis » RACI charts	» Risk analysis » Impact analysis » Force field analysis
MILESTONE	» Set team ground rules and WoW » Confirm have all data » Characterize the current process – have a fully mapped process with 'symptoms' » Stakeholder update (4:30)	» Confirm root cause of issues (detailed problem statements) » Confirm areas of waste/non value add (NVA) » Stakeholder update (4:30)	» Eliminate NVA » Solve root causes » Discrete solution development and alternatives » Stakeholder update (4:30)	» Overall solution collation » New process, new roles and responsibilities defined and detailed » Stakeholder update (4:30)	» Complete solutions » Test solutions (versus business case) » Risk analysis » Implementation issues » Final outbrief to sponsor and stakeholders (3:30)
TIME	8 a.m. to 5 p.m.	8 a.m. to 5 p.m.	8 a.m. to 5 p.m. Potentially late night	8 a.m. to 5 p.m. Team night out	8 a.m. to 5 p.m.

Figure 3-24 The kaizen week

A kaizen can be a very dynamic and inspirational way to design and implement change but it requires strong leadership support and effective communication to those impacted by the changes. The Project Manager understood the risks of this approach and ensured that they were managed within the control plan.

The Project Manager was able to persuade senior stakeholders that this was the appropriate methodology, as previous projects had failed to deliver in part due to inconsistent support from those assigned (conflicting priorities – day job versus project). The agreement on key roadmap stage gates was fundamental to this agreement (Table 3-23). In particular, stage gate 2 was seen as crucial in ensuring that the 13 full-time resources were to be used effectively. The Project Manager would have the authority to cancel or delay the kaizen week if the required data wasn't available or if any team member dropped out at the last minute.

Table 3-23 Project stage gates – purchasing business process improvement

Planning Toolkit – **Roadmap Decision Matrix**				
Project: Purchasing business process improvement			**Sponsor:** Barry Jones	
Date: Week 1			**Project Manager:** Jo Smith	
Stage gate	**Decision**	**Decision by**	**Decision when**	**Data needed**
1	Project Team approval	Sponsor	Prior to project launch	Team selection matrix
2	Go/No go	Project Manager	Prior to kaizen week	Sufficient data and no logistics problems
3	Approval of change	Sponsor	At the end of each kaizen day	List of changes to be approved
4	Approval of kaizen output	Sponsor	At the end of the kaizen week	Kaizen output
5	Project closed	Project Manager	Upon completion of the kaizen action plan	Kaizen action plan
6	Benefits realized	Sponsor	Upon realization of the business case	Benefits realization plan

Scope definition

Once the decision on the project roadmap had been taken the core team considered the fundamental steps to success (Figure 3-25), and also clearer definition of the boundaries of the business process they were to improve (Figure 3-26).

CSF 1 **Accurate and appropriate data** To demonstrate current performance – how well does the purchasing business process work?	CSF 2 **Capable and motivated Project Team** Individuals who currently operate the purchasing business process who are ready and capable of change	CSF 3 **Successful kaizen week** The effective use of the DMAIC process to identify and eliminate waste and to re-design the purchasing value stream
CSF 4 **A capable Project Manager** To use best practice project management to deliver the business change within the cost and time constraints whilst achieving the benefits targets	CSF 5 **An active sponsor and stakeholder group** Leadership and support to ensure that the project has all necessary resources (people, funds, decisions)	CSF 6 **An organization ready for change** Effective communication with those impacted by the business change so that they engage with the change and it is sustained

VISION OF SUCCESS

The sustainable improvement of the purchasing business process so that the divisional budget (for external spend on equipment and services) is in control

Figure 3-25 Kaizen vision of success

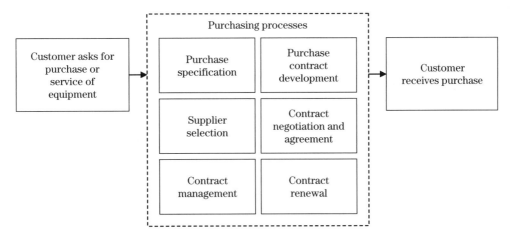

Figure 3-26 The purchasing business process

This enabled them to present and approach an outline scope to the whole team at kick-off so that the team could build the detailed scope. Involvement of the team at the start of planning was crucial in gaining their commitment to activities during the planning stage as they performed their 'day job', and in gaining commitment and understanding of the kaizen week.

The team developed a WBS by firstly brainstorming all the activities required to achieve each CSF. This was then developed into an activity plan (Table 3-24) by agreeing the quantity, quality and functionality of each item in the WBS (activity). Finally the team reviewed the dependencies between items. The activity plan (Table 3-24) was used as the basis for the project schedule development.

Table 3-24 Activity Plan

Planning Toolkit – Activity Plan			
Project: Purchasing business process improvement		**Project Manager:** Jo Smith	
Date: Week 2		**Page:** 2 of 2	
CSF	**Activity**	**Quality, quantity, functionality**	**Dependencies**
1	Data collection plan	All data needed to adequately baseline current performance of the purchasing business process	Core team selected and formed
	Data collection	All data in the plan, accurate and delivered before the kaizen	Team selected and formed
	Data collection tracking	Weekly tracking	Data collection started
2	Team selection criteria	5–10 criteria which are critical to project success	
	Team selection	To match capability needs	Selection criteria needed
	Team kick-off meeting	All team available, used to brainstorm scope and explain kaizen. complete kaizen skill assessment	Team selected
	Team kaizen training	Reduce team capability gap to a pre-defined level	Kaizen skill assessment completed
	Team communications	Weekly email bulletin and core team informal chats during pre-kaizen period	Team kick-off held
3	Kaizen event design	5-day event to deliver full DMAIC process and meet vision of success	Project kick-off Data collection plan
	Kaizen event facilitation	High-level facilitation to support capability gaps, maintain focus and motivation, provide expertize on kaizen tools and the DMAIC process	Kaizen event confirmed and kicked off
	Kaizen logistics plan	A detailed plan of all admin related needs: rooms, materials, food, etc.	Kaizen date confirmed
	Kaizen logistics	Delivery of the logistics plan	Plan developed

(continued)

Table 3-24 (Continued)

CSF	Activity	Quality, quantity, functionality	Dependencies
		Planning Toolkit – Activity Plan	
Project: Purchasing business process improvement		**Project Manager:** Jo Smith	
Date: Week 2		**Page:** 2 of 2	
3	Kaizen readiness checklist	A list of acceptance criteria to ensure that the kaizen has the highest probability of success – if they are not achieved then the kaizen event will be delayed or cancelled	Understand any data gaps, logistic challenges and stakeholder and target resistance
	Kaizen readiness tracking	A daily review of acceptance criteria 2 weeks prior to the scheduled event	Agree acceptance criteria within core team
	Kaizen success criteria	A list of criteria which the kaizen must meet by the end of the event	Agree success criteria with sponsor
	Kaizen success criteria tracking	A daily review of the success criteria during the event	Kaizen event
4	Project delivery planning	A detailed plan of how, who, when	Project kick-off
	Project action tracking	A 'live' action list generated from core team meetings	Core team weekly meetings pre-delivery
	Project risk assessment	A 'live' log of potential threats and opportunities, mitigation and contingency plans	Core team weekly meetings pre-delivery
5	Sponsor contract management	Regular contact with the sponsor pre-delivery to ensure his active support	Agreed sponsor contract
	Stakeholder management plan	A plan to target key stakeholder groups so that there is sufficient engagement for sustainability of the new process	Stakeholder analysis
	Communications plan	A plan of key messages to key audiences during all project phases	Stakeholder management plan
	Stakeholder management	Delivery of the stakeholder management and communications plans	Stakeholder management and communications plans
6	Culture readiness survey	A questionnaire which would baseline resistance and support for change	Before event
	Kaizen outbrief invitations	Daily event briefings to all those currently working in the business process and all senior stakeholders	Before event
	Communications event	The full communication of the new process to all those currently working in the business process – a training event	After event

The team were then able to develop an RACI Chart to agree who would be involved in each activity:

➤ Responsible – who will actually perform the activity.
➤ Accountable – who is accountable for the successful completion of the activity.
➤ Consulted – who needs to be consulted in order for the activity to be successfully completed.
➤ Informed – who needs to be informed of the outcome of the activity.

The data collection plan was also presented to the team so that they could agree how data could be collected during their 'day jobs'.

The activity plan (Table 3-24) helped the team to more fully understand the scale of the project and their role within it. The data collection plan introduced them to a different way of measuring performance: introducing cycle time, purchase processing models and management efficiency, for example, which also helped to define the scope of the project.

Funding strategy

The sponsor was able to fund the project through the divisional expense budget and required the Project Manager to follow usual approval processes for this funding:

➤ Resource plan and indication of any loss to the business through not having the resource available for BAU.
➤ Expenses plan mainly linked to the logistics of running a week-long workshop.

Once approved, through a system of allocating approved funds to different cost codes, the Project Manager had authority to commit funding in line with the plan.

Conclusions

The project delivery took 3 weeks as scheduled and the planning took 6 weeks. The AAR identified effective set-up planning as one of the keys to success in delivery and also one of the main reasons that each kaizen team member enjoyed the experience so much.

Key points

The aim of using this particular case study was to demonstrate that:

➤ For small, fast-track projects, the planning can typically take longer than the actual project delivery. Other examples of this include shut-down projects, new product or service launch projects, IT system changeover projects, etc.
➤ Getting the right mix of capabilities within the Project Team is critical to project success.
➤ A project can take many different routes and a good Project Manager will define the direction and the key decision points along the journey – this is the value of a project roadmap.

Handy hints

Use organizational resources wisely

This might seem obvious but any Project Manager needs to realize that when organizational resources (e.g. money, people) are allocated to his project they cannot be used elsewhere. In effect any misuse by you represents a waste of resource efficiency to the business which can never be re-couped.

Be sure that you are the 'right' Project Manager for the job

Sometimes organizations inappropriately assign Project Managers to projects, usually because they do so on a 'who's free' basis rather than trying to fit a Project Manager to a specific business challenge. Be sure that you have the capability, the capability potential and/or the support to deliver what is expected of you.

Team building is a necessary part of project management but do it wisely

Team building isn't just about playing games together. It has to be about building a shared commitment to a goal, a vision of success. Ensure that your team building activities are focused and relevant as the team can soon get fed up of pointless fun.

No matter how simple the project or how similar it may be to a previous project, define a project roadmap

A roadmap is a communication tool for everyone involved in a project. It tells us where we are going and what we have to do to move through successive stages of the journey.

There are many ways to define scope, however, to be sure that it is complete and appropriately uses the path of CSFs

If a path of CSFs is appropriately defined, it can be used effectively in the detailed scope definition, acting as a checklist to whether the scope is of sufficient quantity, quality and functionality. Remember CSF achievement is critical to achievement of the project vision.

Don't ignore funding strategy – there are organizational and legal implications

Organizations usually have very strict rules and guidelines on how money can be approved and then actually spent. This is usually linked to their own business governance as well as accounting law. Find out the rules and make sure that the team understand them.

Setting up a project successfully needs people – launch the project

A key part of engaging the team in the project and in delivering the shared goal is getting the team involved in the detailed planning early. A Project Manager launching a team too early can run the risk of appearing confused and disorganized. A Project Manager launching a team too late can appear disinterested in team engagement.

And finally . . .

A robust project set-up plan:

- Articulates the project set-up needs in terms of who and what will be delivered. It considers the set-up of the organization's resources (e.g. people, money).

- Has a link to the project business plan, ensuring that the business benefits will be enabled.

- Has a link to the project control plan and its potential to deliver the project in control.

- Is built on sound relationships and effective Project Team management.

4 Control planning

In the context of project delivery planning, control planning is the way that we define how the project is to be controlled. It is based on the approved project business case which has, or will be, developed into a business plan (Chapter 2). It is also based on a set-up plan (Chapter 3) which defines the project scope and how it is to be organized and administered.

The control plan continues to support the development and management of the relationship between the Project Manager and the Project Team. If they have shared and agreed control processes they are more likely to deliver a successful project outcome.

The project delivery plan (PDP) must provide an appropriate baseline from which to deliver the project and assess progress during delivery (Figure 4-1) and this is the main goal of a control plan.

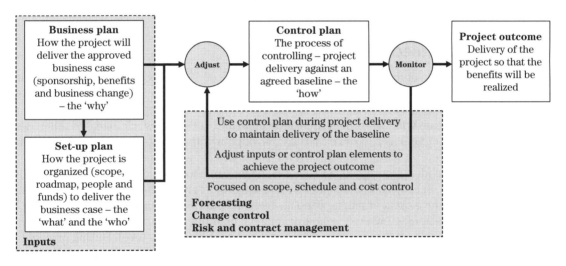

Figure 4-1 Using a control plan

Historically this has been the strongest part of a PDP. This is because it is focused around the traditional project management control tools:

- **Schedule development:** Estimating the time it takes the project to be delivered; setting the schedule baseline.
- **Cost plan development:** Estimating the total project cost; setting the cost baseline.
- **Risk plan development:** Forecasting likely risk scenarios and developing mitigation and contingency plans.
- **Contract plan development:** Developing a plan to optimize use of external resources.
- **Change control:** Setting up processes to control the delivery of the project scope to enable the realization of the business benefits.

However, the set up of these tools is only a part of the control plan. To be effective the tools need to be used in an environment of appropriate monitoring, forecasting and then adjustment. Controlling a project relies on monitoring current progress versus baseline plans and reviewing external drivers that can impact project success.

A control plan aims to define this methodology for a project and this supports the working relationship between the Project Manager and the Project Team. Projects can fail when these essential processes, tools and methodologies are missing or are not effectively managed.

What is a project control plan?

Within a PDP, a project control plan is the formal articulation of 'how' the project will be managed and controlled so that the business is assured of certainty of outcome regarding the original business case. It covers the following four planning themes:

- Risk management strategy.
- Contract and supplier strategy.
- Project control strategy.
- Project review strategy.

This part of the PDP defines 'how' the project will be delivered, 'how' risks and external resources will be managed and 'how' the performance of the project will be assessed.

How to develop a project control plan?

As discussed during project set-up planning (Chapter 3), the role of the Project Manager is to facilitate the development of the PDP and to decide on the appropriate style of facilitation for each planning theme.

The control plan is essentially a tool to plan, monitor and forecast project activities and therefore it is a tool which Project Team will use to plan their individual and sub-team work load. On this basis it is appropriate for the Project Manager to develop the control plan with strong participation from team members.

Using the facilitation model (Figure 4-2), a Project Manager can choose one of four different facilitation styles. Selection depends on Project Team capability and the project situation or planning theme being developed. Control plan development can be used as a way to 'contract' with and between Project Team members; clarifying roles in meeting cost, schedule, quality, quantity and functionality acceptance criteria. It is therefore important that team involvement is at an appropriate level. During project planning the Project Manager usually takes on a facilitator role in order to ensure optimum team involvement:

- **A facilitator:** Someone who contributes structure and process to interactions so groups are able to function to achieve required outputs.
- **Facilitation:** A method of providing leadership without taking responsibility and ownership from the group who required the specified outputs.

It is not unusual for the Project Manager to involve an external facilitator as he may need to be involved in some of the content development and also in the ownership post-development. Whoever takes on this role will need to do the following to be successful:

- Select the most appropriate facilitation mode (Figure 4-2) for a specific session.
- Define and then achieve the defined success criteria for the session.

		CONTENT MANAGER	INFLUENCER
Degree of intervention	HIGH	Project Manager leads the Project Team around a traditional process (e.g., brainstorm, categorize, rank) around a flip chart and strongly influences direction and content	Project Manager leads the Project Team around a creative, high-risk process, designed specifically for the group/activity. Likely to use novel activities and deliver a broad spectrum of possible outcomes
		PROCESS MANAGER	GAMES MASTER
	LOW	Project Manager introduces a process to the Project Team with defined times and output targets and then lets them get on with it, only monitoring the process in terms of time and content	The Project Manager introduces a novel game/process and allows the Project Team to find their way through it, only intervening if direction is requested
		LOW	HIGH
		Level of innovation	

Figure 4-2 Potential facilitation modes, for control planning

Facilitation modes

The facilitation model (Figure 4-2) describes two interacting criteria:

- **Level of intervention:** This describes the amount of involvement and direction from the Project Manager or external facilitator, either high or low. An indication of this is the percentage of time the facilitator spends talking rather than the group being facilitated.
- **Level of innovation:** This describes the activity being used as the basis for developing a planning theme, either a very novel one or standard. The more customized a process the more innovation has been used to design the session.

The result of having all possible combinations of these criteria is four different modes of facilitation.

Content manager

For most Project Managers this is the normal mode when working with a Project Team. The Project Manager introduces the process, which the group will follow for a particular meeting, and is involved in the development of the final outputs.

For example, during a project planning meeting to review external resource selection, a Project Manager would highlight the project process which is required to select external companies. As different parts of the scope are bounded into procurement packages the Project Manager will get involved in defining selection criteria. He will be highly involved in the meeting and as the chair will strongly influence the outcome through direct questioning and responding to questions from team members.

Process manager

Project Managers recognize that some situations do require less hands-on involvement from themselves, or an external facilitator. This model can be used when a Project Team needs to be heavily involved but within a fairly standard process that they have used before.

For example, during a project risk mitigation planning session, the Project Manager can introduce a standard risk process and then leave the team to identify risks, describe the consequences and then score the risk in an objective way (the scoring system is usually defined beforehand). In this case the Project Manager will be doing less talking and more directing.

Influencer

Sometimes a project situation requires some innovation if the appropriate solution is to be identified. This usually means that the session design is a larger component of the whole facilitation process, and usually involves customized development. Customization may be anything from using two standard tools together to developing a whole new tool or process.

For example, a Project Team needed to develop a suitable project scorecard from which they could succinctly report progress internally. The customers of the scorecard were all Project Team members although only a part of the team were to be involved in developing the tool. The Project Manager needed to be involved so that his views on project reporting and the importance of certain measures could be taken into account. The Project Manager defined a process which started with brainstorming 'how do we know the project is not performing?' The ideas were then categorized and each team member rated them in terms of 'how many times I have seen this in previous projects'. The top 8 were converted into 8 measures of project success on the basis that if this isn't happening then we're on track. This was a highly innovative process which required the Project Manager to direct the teams very closely. The result was an appropriate customized scorecard which the team owned.

Games master

Within a project planning scenario it is less likely that a Project Manager would need to use a high innovation, low intervention process. This tends to be a facilitation mode used during project delivery when teams are in a solution development or implementation phase.

For example, a Project Team training a part of the business in a new process decide to use role playing to test new standard operating procedures (SOPs). The Project Manager ensures that the 'game' is set up appropriately and then stands back to watch the game being played. If the SOP is robust then it will be capable of handing the simulated situations and the business users will have been trained in a safe (and memorable) environment.

Facilitation process

A good meeting of any type requires some element of design. An appropriate tool to formulate the design is an input-process-output (IPO) diagram (page 2):

- *Outputs* – consider the outputs needed from the meeting. For example, is the goal that the overall project schedule will be defined by the end of the session?
- *Inputs* – review what data and resources are available. For example a work breakdown structure (WBS) may have been developed already and a Project Team may be in place.

- *Process* – based on the inputs available and the outputs required a facilitator should be able to design a session to convert one to the other. For example, a schedule development 'brown paper' exercise may be used to maximize team involvement (and ultimately ownership in the resultant schedule). This type of exercise is based on the team building a manual logic-linked, resource loaded schedule on a large piece of brown paper which usually covers one wall of a meeting room. The Project Manager can step back because he is happy that the WBS is complete and robust. He knows that the team are capable of identifying dependencies and task durations based on their previous experiences.

A Project Manager has been successful if:

- He has an agreed IPO which meets the needs of the project.
- He has an appropriate session design which, combined with the selected facilitation mode, aims to meet the required session objectives.
- He achieves the required session objectives.
- The team functions effectively during the session.

In terms of success criteria for the actual session, some hints for the facilitation 'on the day' are:

- Start with a clear understanding of the process – use the IPO to develop a shared understanding of what the group are aiming to achieve in the session and how they are to do that.
- Treat all team members as equals – successful facilitation is about motivating team members to provide you with their knowledge, experience and perspective.
- Develop clear outputs (interim deliverables) which are owned by the team members.
- End with a summary of the session 'journey', the achievements and the success of the team in terms of their participation.
- During the facilitation process (whatever the mode):
  - *Stay neutral – allow the process to support any discussions or disagreements*. If the Project Manager has to jump in and decide a point then this can disempower the team.

    For example if one team member thinks a risk is high and another thinks a risk is low then the Project Manager as a good facilitator will remind the team members of the scoring system, will ask questions so that the consequence of the risk can be further discussed and would direct the team to follow the process. He would not just state 'it's obvious it's high' unless directly asked to do so by the team. In that situation the Project Manager should take the team through the process to get to an objective decision.
  - *Listen actively, manage discussions and give clear summaries.*

    For example, if one team member is taking over a session then the Project Manager should use his facilitation skills to bring in other team members. In addition he should summarize ideas and check for team buy-in, particularly if a session has a few noisy participants and a lot of quiet ones.
  - *Ask questions, paraphrase and synthesize ideas.* A Project Manager should be helping the team to move forward at an appropriate pace, and often some focused questioning 'do you mean x', 'does that mean y', is appropriate.
  - *Give and receive feedback.* Particularly with unfamiliar processes, a good facilitator should be letting the team know that the process is going to plan, or that an idea was a particularly good one or that a suggestion was innovative but not within scope.

In terms of control planning, all four planning themes will require significant input from the team and a Project Manager needs to be a good facilitator to harness their ideas, experience, knowledge and skills. In some respects this is a critical component in the development of a control plan.

Risk management strategy

In some respects the main goal of a control plan is to control the project in order to maximize the certainty of outcome. Therefore a Project Manager needs to define how uncertainty within a project will be managed. During a project there are likely to be scenarios where potential threats or opportunities may impact the project outcome. We usually refer to these as risks and we manage them by:

➤ Identifying and analyzing all possible risk scenarios.
➤ Developing and managing of appropriate risk responses to take advantage of potential opportunities and mitigate potential threats.

The control plan needs to define a risk management strategy for a project which usually would cover:

➤ Risk management process.
➤ Scenario planning.
➤ Risk management tools.

Risk management process

The first step in defining a risk management strategy is defining the process to be used, when and how (Figure 4-3).

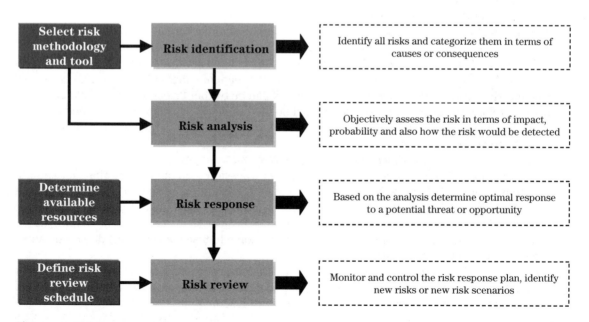

Figure 4-3 Risk management process

Select risk methodology and tool

It is usual to define the overall methodology, tool or set of tools which will be used to assess risk on a project. No matter the choice of tool, the process from this point onwards remains the same. Methodology/tool selection is discussed later in this chapter and highlights the reasons why one tool might be more appropriate than another in a specific project situation. On any one project a manager may have a selection of tools in use to support the management of different types of risk.

Risk identification

The most critical step in developing a risk management strategy is to identify all possible scenarios where uncertainty exists. Scenario planning is discussed later in this chapter and highlights the various methodologies to identify these situations (page 152).

It is usual to categorize risks once identified and this can be done a number of ways:

- By risk scenario:
    - Project stage, such as all risks related to the design stage.
    - Failure mode, such as all risks likely as a result of an incorrect design decision.
- By risk consequence:
    - Specific project objective, such as risks with a cost or schedule impact.
    - Critical success factor (CSF), such as all risks impacting the achievement of CSF 1.

Risk should be categorized in a way which supports the next stages in the management process. For example, it is common to collate all risks which could impact cost so that cost contingency can be developed, managed and controlled (page 187). Different risk tools will direct you to categorize risks in different ways to support the specific methodology in use.

Risk analysis

The aim of any risk analysis is to be as objective as possible:

- How likely is it that the risk will occur?
- What is the impact on the project?
- Will the project detect if this risk becomes an issue?

Not all risk tools 'score' each of these parameters, although most will assess impact and probability so that some ranking of risks can be defined.

Determine available resources

Within any project there will always be a finite level of resource (time, money, people, assets, materials) and it is useful to review the resources available before developing extensive plans to mitigate all risks. It is not practical to aim for this goal nor is it necessary.

Risk response

Based on the priority associated with a risk, three types of response are likely:

- *Mitigation plan A* – an action to reduce the probability of a risk occurring.
- *Mitigation plan B* – an action to reduce the impact of a risk should it occur.
- *Contingency plan* – an action to deal with the risk as it occurs.

It is usual to only develop mitigation and contingency plans for risks scored as 'high' and then mitigation plans for 'medium' ranked risks. Contingency plans are not usually implemented until it is clear that the associated mitigation plan is not successful.

Define risk review schedule

A risk management strategy should also be clear on the timing and need for risk assessment reviews and/or other types of risk assessment. Typically these would be linked to specific project stage gates (Chapter 3). For example, a review of a facility's constructability might be conducted during conceptual and front end design as well as at the end of detailed design.

Risk review

The risk review process is continuous. It usually commences during business case development, before being completed more formally during PDP development, and is then used as a part of the control strategy during project delivery.

Risks change as the project is being delivered. Some risks never occur whilst others quickly turn into issues requiring contingency actions. New risks can also occur and these need to be identified as soon as possible so that appropriate action can be taken. Ultimately, a risk review summarizes the degree of certainty of outcome, and whether it is positive or negative.

Scenario planning

A project environment has many areas of uncertainty and a Project Manager needs to ensure that as far as possible all potential risk scenarios are identified during the planning stage. There are many ways to do this and three typical ones are presented:

- Risk brainstorming.
- What if analysis.
- Scenario development.

Each relies on using the Project Team to evaluate what might happen during the project, focussing on both internally generated risks and those caused by external drivers outside of the control of the team.

Risk brainstorming

A classic way to identify risks to project success is to use a Project Team session and have the Project Manager facilitate a brainstorming session. The key to a good brainstorm is to ask a clear, succinct question which is easily and consistently understood by all. In the case of a risk brainstorm it is easy to just ask a team 'what could go wrong?' There are various methods to organize and summarize a brainstorm but the one demonstrated here is an affinity diagram (Figure 4-4). In this example the team have brainstormed all the things which they think will make a relocation project go wrong and then they have grouped together all similar items and given the grouping a category name. The categories are the main risk types which would be analyzed.

What if analysis

Another way to get a team to think differently is to use a structured brainstorm technique based on a 'what if' analysis. This methodology helps to build realistic scenarios against a specific 'what if' statement,

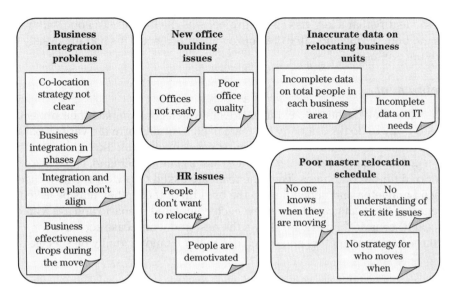

Figure 4-4 Example risk affinity diagram

by looking at the worst case scenario and the scenarios leading to this. It not only identifies risks but starts to highlight those which will more than likely lead to the worst case. For example, in a process improvement project the team brainstorm what they perceive as the worst case: 'what if the process isn't improved by the project?' only to discover that this risk could lead to a much worse situation when they continue to ask 'what if?' to each outcome of the previous question (Figure 4-5). This is a different way to look at risk identification because it does focus on consequences. In the example the Project

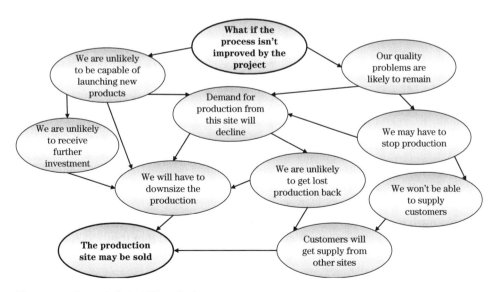

Figure 4-5 Example 'what if' analysis

Team were able to identify that a key risk to the project was the inability to cope with required production demand both during and after the improvement project. This had previously not been an identified area of uncertainty.

Scenario development

At times a team need to look further into external drivers which could impact their project and a scenario brainstorm is a good way to do this. Initially the team need to brainstorm the various parameters in each scenario, and then to identify what the best and worst cases might look like.

For example, a product launch team might want to assess launch demand, capacity or product quality against a number of different outcomes. Then the team are asked to review the best case and worst case scenarios for each parameter, and depending on the breadth of these, they may be asked to review some intermediate scenarios. The team can then review each completed scenario and ask 'can our project still be successful in each scenario?' Figure 4-6 shows the best and worst case scenario for a product launch project against three basic parameters and concludes that it cannot actually be successful in either scenario.

Scenario parameter	Scenario A – 'best case'	Scenario B – 'worst case'
Launch demand	Demand exceeds current forecasts	Demand is less than half of demand forecasts
Production capacity	Production capacity meets design	Production capacity is only 75% of design
Product quality	Reject rate is <1%	Reject rate is >10%
Production staff	Highly capable team driving continuous improvement	Demotivated team waiting for staff reductions
Lifecycle costs	Facility lifecycle costs are less than budget	Facility lifecycle costs exceed budget
Facility capital cost	Facility capital cost 10% under approved funds	Facility capital cost >10% over approved funds
Project outcome	Production is highly efficient but still disappoint the market by not having sufficient launch stock	Launch stock can be met but ongoing supplies to the market are erratic and profit to the company is reduced by waste and high operating costs

Figure 4-6 Example scenario development

In this case two further scenarios were completed with less extreme outcomes. These showed that the project was extremely sensitive to both demand and quality and that they did represent areas of significant uncertainty. The scenarios also highlighted to the team the importance of lifecycle costing in maintaining company profit levels.

Tool: Project Scenario Tool

The aim of this tool (Table 4-1) is to support Project Teams in developing scenarios through structuring the brainstorm and directing the scenario development process.

Table 4-1 Project Scenario Tool explained

Planning Toolkit – **Project Scenario Tool**				
Project: <insert project title>		**Sponsor:** <insert name>		
Date: <insert date>		**Project Manager:** <insert name>		
Scenario parameter	**Scenario**			
	Scenario 1	**Scenario 1**	**Scenario 1**	**Scenario 1**
<insert parameter>	<insert a description of the parameter in this scenario>	<insert a description of the parameter in this scenario>	<insert a description of the parameter in this scenario>	<insert a description of the parameter in this scenario>
<insert parameter>				
<insert parameter>				
Project outcome	<insert how the project would cope in this scenario>	<insert how the project would cope in this scenario>	<insert how the project would cope in this scenario>	<insert how the project would cope in this scenario>
Summary				
<insert any summary comments regarding project outcomes in terms of the most likely scenario and whether further mitigating actions are needed in order for the project to be successful within this scenario>				

Scenario parameter

This is a description of the parameter which will change in each scenario. It is best if this is succinct and is clearly a single parameter. For example, this could be the state of an external driver outside of the control of the project. Typically the scenario parameters are areas of high uncertainty. Sometimes the analysis can show something isn't as uncertain or as variable as first thought, for example, if it doesn't change much between scenarios.

Example parameters for a product launch project are shown in Figure 4-6. Other examples of parameters which might be used within a business change project are:

➡ The current state of the external environment or a specific industry sector.
➡ Customer behaviours.
➡ How the business supports the changes.
➡ The availability for capital investment in the business.

Generally scenario analysis is most useful when the scenarios are built from a selection of parameters which are generally outside of the control of the project and the Project Manager.

Scenario

It is often useful to give a scenario a name or title, such as 'worst case', 'ideal world' or 'downsized site', which gives an indication of the overall flavour of the scenario. It is important that all scenarios are possible given the right set of circumstances, however it is likely that scenarios 1 and 4 would represent extremes (positive and negative). For each scenario a parameter should be considered and a view of what that parameter would be like needs to be developed. However, the scenario should have consistency between parameters and above all else, be realistic. For example, is it likely that 'ideal world' for a production plant would have poor quality coupled with demand exceeding capacity?

In Figure 4-6 each parameter has been reviewed in two opposing scenarios: 'best case' and 'worst case' for a product launch project. Each complete scenario is internally consistent:

- It is likely that both the designed capacity is achieved as well as the quality targets.
- It is likely that the team could be demotivated by the level of ongoing change and that in this situation quality will suffer.

However, it is also important that less extreme scenarios are developed to test how sensitive the project outcome is to each scenario and each parameter.

Project outcome

Finally, there needs to be a review of how the project will be impacted by each scenario. A 'what if' exercise is usually conducted and some summary outcomes should be detailed.

Summary

It is important to review the scenarios developed, their likely impact on the project outcome, and then to conclude with what actions are needed to mitigate the risks identified. The aim is that the project can still be successful within a number of different scenarios bearing in mind that most parameters are outside of the control of the project. The project cannot control which scenario, or combination of scenarios, will actually happen, only how the project can react to it.

Risk management tools

There are many tools to support the development and deployment of a risk management strategy and a Project Manager needs to select the appropriate mix of risk tools for a specific project (Table 4-2).

Table 4-2 Examples of available risk tools

Risk tool	Description	Reason to use	Comments
1. SWOT analysis	To identify and analyze the project strengths, weaknesses, opportunities and threats so that an appropriate action plan can be put in place	➤ A good overall review of a project situation ➤ Generates a high-level plan to address key issues	A tool template is included in this section with explanation on 'how to use' (Table 4-4)
2. Risk table and matrix	To identify risks to achievement of project CSFs and, through prioritization, the identification of appropriate mitigation and contingency plans Can also be used on a WBS and categorization other than CSF	➤ To assess the likelihood of overall project success ➤ Identifies CSF or other categories within the project scope where risks are highest ➤ Supports resource management through prioritization	A tool template is included in Appendices 9-12 and 9-13 with explanation on 'how to use' this section
3. Critical path of risks	This tool relies on using the risk table and matrix and is effectively an 'add on' visual tool to show a risk profile versus the CSFs	➤ To assess the likelihood of overall project success ➤ To identify CSFs that will/will not be achieved	Tool is discussed in this section
4. FMEA (Failure Mode Effect Analysis)	This tool is a methodology focused on the elimination of defects based on a risk priority score	➤ To assess and correct individual parts of a process, system or product ➤ To highlight the most critical defects	Tool is discussed in this section
5. Risk flowcharts	This tool is based around a decision tree methodology asking risk-related questions and determining action based on the response	➤ To quickly identify high-risk areas ➤ To identify critical actions for risk reduction	Tool is discussed in this section
6. Risk checklists	This tool is based around a keyword methodology asking risk-related questions against key risk areas	➤ To identify specific types of risk such as safety or compliance ➤ To identify critical actions for risk reduction	Tool is discussed in this section
7. Scope risk assessment	This tool uses the basic risk process against a specific consequence category (scope)	➤ To assess risks to the quality, quantity and functionality of project scope	Tool discussed in Chapter 3 (page 128)
8. Cost risk assessment	This tool uses the basic risk process against a specific consequence category (cost)	➤ To assess risks to the achievement of the cost plan	Tool discussed in cost planning section (page 187)
9. Schedule risk assessment	This tool uses the basic risk process against a specific consequence category (time)	➤ To assess risks to the achievement of the baseline schedule	Tool discussed in schedule planning section (page 195)

SWOT analysis

SWOT stands for strengths, weaknesses, opportunities and threats. The SWOT methodology is a very well-known organizational analysis tool. It supports an organization in identifying what it is good at and in scanning the external environment for things which may impact it in a positive or negative way. To complete the analysis an organization then selects which combination of SWOTs to work on that will be the best use of resources in order to deliver the optimal outcome.

A project-based SWOT analysis is an adaptation of the organizational SWOT. It allows the Project Team to conduct an objective internal review and then to scan the environment external to the project (usually within the organization or the industry) for potential opportunities and threats. SWOTs should be ranked in terms of their importance to the project outcome, then those with the highest ranking should be converted into a resourced action plan. Without this latter activity the SWOT analysis is unlikely to support appropriate project development and merely becomes a project audit.

For example, a project to upgrade a part of an existing food processing facility conducted a project SWOT analysis as a part of early planning (Figure 4-7). Many strengths and weaknesses were identified and some high-level opportunities and threats.

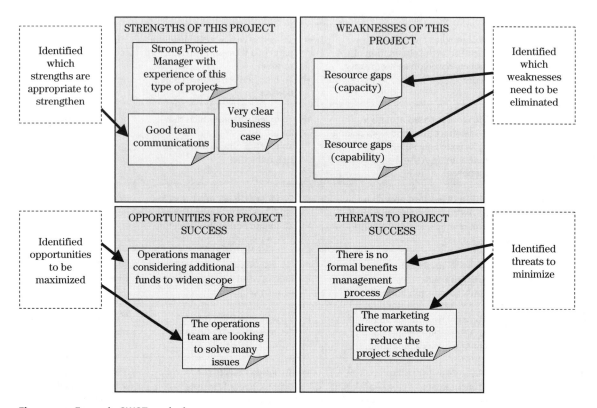

Figure 4-7 Example SWOT analysis

Figure 4-7 shows a part of this analysis, from which the team selected items for action:

> *Strength* – team communications were already good but needed to be strengthened to ensure that this was never an issue within the project.
> *Weaknesses* – the team recognized that they didn't have the complete skill profile to perform some of the work, nor did they have enough people to do the work in the time required. This was a weakness which needed to be eliminated if the project was to achieve its objectives.
> *Opportunity* – the team recognized that there were opportunities to maximize success by meeting key stakeholder goals.
> *Threat* – the two threats had quite different consequences for the project. The greater of the two was the pressure the business was under to complete the project sooner. The Project Manager needed to eliminate this threat, recognizing that with current resource gaps the current schedule was barely achievable.

A clear action plan was then put in place against each of the selected SWOT areas.

Risk table and matrix

The Risk Table and Matrix (Melton, 2007) are simple but effective methods of identifying and managing risks to project success (Appendices 9-12 and 9-13). This methodology relies on the identification of risks, then assessment of probability and impact, and assumes that all risks are detectable. The consequences are considered in terms of the impact on either overall project success or specific CSF success.

The use of the tool requires some basic pre-work:

> Decide how the risks are to be categorized – by CSF or using the WBS.
> Define a scoring system (Appendix 9-13).

Once the pre-work has been completed, tool use follows three steps (Figure 4-8):

> *Step 1*: Identify the risk, describe how it occurs and the consequence to a CSF or objective. Score the risk as low, medium or high based on its probability and impact score.
> *Step 2*: Plot each risk on the matrix. This is a very visual prioritization tool: 'green risks' are low priorities and mitigation plans may not be resourced, whereas 'red risks' are high priorities and mitigation plans should be developed, resourced and tracked.
> *Step 3*: Mitigation and contingency plans are developed based on priority.

Critical path of risks

One of the benefits of using the path of CSFs as the categorization process for the Risk Table and Matrix is that the results can be displayed as a critical path of risks (Figure 4-9). Apart from being a very visual method to demonstrate which CSFs are at risk, it can also give an overall assessment of the likelihood for project success. An example of this is shown in Case study 3 (Figure 8-13).

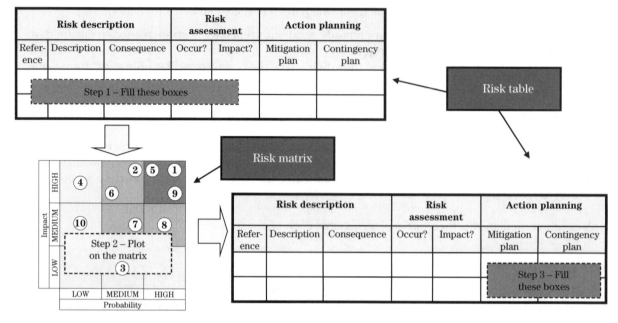

Figure 4-8 Using the Risk Table and Matrix

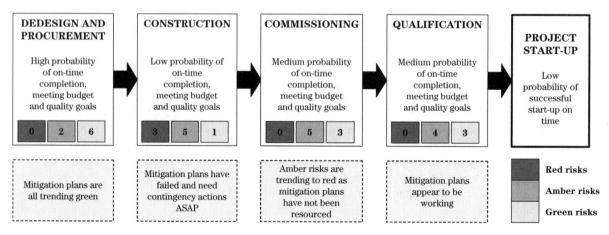

Figure 4-9 Example critical path of risks

It can also be used with a more traditional WBS based around a project roadmap stage definition (Figure 4-9), where each phase is actually connected via the project schedule through a more traditional critical path analysis (CPA). For example, a pharmaceutical facility project is being built and the risk assessment has been categorized against the four key stages (design and procurement, construction, commissioning and qualification). For the project to be a success each must be completed successfully (Figure 4-9).

The benefit of this methodology is that it allows the Project Manager and Project Team to focus on overall success rather than just one high-risk area. This allows a broad range of mitigation plans to be developed, moving the whole project forward in its management of uncertainty.

FMEA

FMEA stands for 'failure mode and effect analysis'. It is a structured method for identifying and analyzing failure modes/defects within a process, system or product. It is usually completed on a component-by-component basis and produces a risk priority number (RPN) that allows prioritization of any subsequent action planning.

Phase 1

The initial part of the process involves failure mode or defect identification, description and then scoring (Table 4-3):

➤ **Failure mode:** For each process step or system/product component brainstorm the various potential failure modes or possible defects (things that could go wrong). For example, in a product design project a failure mode during early design stages could be selection of incorrect materials.
➤ **Failure effect:** For each identified failure mode describe the consequence. For example, in our design project, the ultimate consequence of an incorrect material specification could be that the product either does not perform or is not safe to perform.

Table 4-3 Example FMEA

Phase one									Phase two				
Failure mode	Failure effect	S E V	Causes	O C C	Controls	D E T	R P N	Action plan	P S	P O	P D	P R P N
Missing information	Search for information	6	Human error	2	Entry check	1	12	Kanban system to control flow of information	6	1	1	6
Poor information	Search for information	4	Human error	2	Quarterly VAT review	6	48		4	1	1	4
Filing error	Search for information	6	Human error	3	End of year audit	10	180		6	1	1	6
Data entry error	Re-entry	4	Multiple entry	5	End of month check	4	80	New accounting IT system	4	1	4	16
Accounting error	Audit	8	Misaligned systems	3	End of year audit	10	240		8	1	10	80

Scoring system		
SEV: 1–10 based on time to correct	**OCC:** 1–10 based on likely frequency of occurrence	**DET:** 1–10 based on time to detect

- **Severity (SEV) score:** A numerical score is assigned based on the failure effect. The scoring system should be defined before the start of the FMEA analysis (Table 4-3).
- **Causes**: Explain the cause for each identified failure mode.
- **Occurrence (OCC) score:** A numerical score is assigned based on how likely the cause would be to occur. The scoring system should be defined before the start of the FMEA analysis (Table 4-3).
- **Controls**: Identify the current controls in place which would detect the failure mode.
- **Detection (DET) score:** A numerical score is assigned based on the controls and the ability for the failure mode to go undetected. The scoring system should be defined before the start of the FMEA analysis (Table 4-3).
- **RPN:** A risk priority number is calculated based on previous scores: RPN = SEV × OCC × DET.

At this stage a Project Team would be able to rank the failure modes according to RPN score. The example in Table 4-3 is for an accounting system improvement project. Here, the FMEA is used as a design tool looking at the risks in the current accounting process.

Phase 2

After the RPN has been calculated the second part of the analysis can begin:

- *Action plan:* Based on the RPN define which failure modes require action and then define the action plan.
- *Rescoring:* Based on the action plan review whether the severity (PS), occurrence (PO) and detection (PD) score has or will alter and therefore calculate an updated risk priority number (PRPN).

The reduction in RPN will be an indication of the reduction in risk. For example, in the accounting system improvement project, the FMEA analysis (Table 4-3) was able to identify the positive impact of key areas of design through significant reduction in all RPNs.

Risk flowcharts

A structured methodology to address a specific risk in different areas of a project, process or system is to use targeted risk questioning alongside a defined decision tree. This will indicate what action should be taken in order to mitigate the risk and it relies on:

- A clear set of risk questions which will identify if a risk scenario is present. This requires the ability for a digital response, for example yes or no.
- A logical decision tree which can take the response to the risk question and either ask a further risk question or propose an appropriate and specific action.

For example, a team involved in a project to refurbish a pharmaceutical facility is defining the scope and need to ensure that it addresses all current compliance issues. A risk flowchart (Figure 4-10) was used to review each area of the existing facility with only two possible outcomes:

1. Include in scope (any action to eliminate a compliance risk or to redesign so that the risk is mitigated).
2. Exclude from scope (any area which is in compliance).

Not only did this process highlight areas of current compliance but it clarified the areas of the facility which did not require any refurbishment. The latter allowed significant scope reduction from early estimates, and the value-add expenditure of funds completely focused on risk reduction (for the business and for the customers of the business).

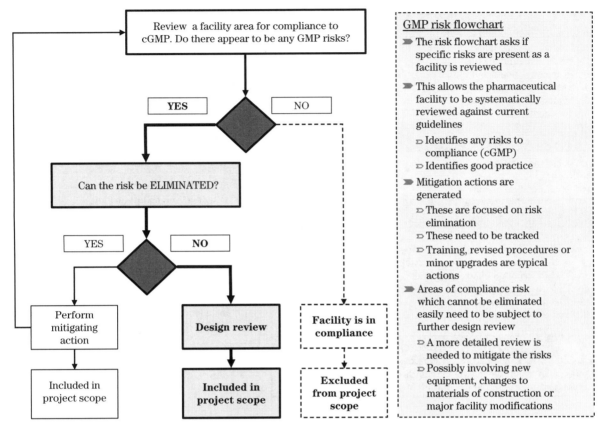

Figure 4-10 Example risk flowchart

In Figure 4-10, the risk assessment was purely to look at quality risks, however flowcharts of this type of tool can be used for other categories of risk and can become much more complex.

Risk checklists

Another simple but effective way of organizing risks and developing mitigating actions is a risk checklist. This tool uses a set of predefined questions or guidewords which can highlight if a risk (opportunity or threat) is present or not. Typically, these are used on a specific risk consequence category such as quality, safety, operability and design effectiveness.

For example, within a food manufacturing facility the design of a new unit needs to be checked for any quality risks linked to compliance to current Good Manufacturing Practices (cGMP). Three types of quality risk have been identified (Figure 4-11):

- Risks to quality from the facility environment.
- Risks to quality from the equipment and internal processing environment.
- Risk to quality from the way the product is handled.

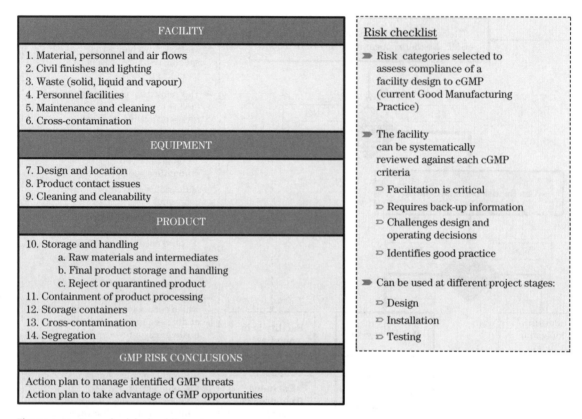

FACILITY

1. Material, personnel and air flows
2. Civil finishes and lighting
3. Waste (solid, liquid and vapour)
4. Personnel facilities
5. Maintenance and cleaning
6. Cross-contamination

EQUIPMENT

7. Design and location
8. Product contact issues
9. Cleaning and cleanability

PRODUCT

10. Storage and handling
    a. Raw materials and intermediates
    b. Final product storage and handling
    c. Reject or quarantined product
11. Containment of product processing
12. Storage containers
13. Cross-contamination
14. Segregation

GMP RISK CONCLUSIONS

Action plan to manage identified GMP threats
Action plan to take advantage of GMP opportunities

Risk checklist

➤ Risk categories selected to assess compliance of a facility design to cGMP (current Good Manufacturing Practice)

➤ The facility can be systematically reviewed against each cGMP criteria

  ▷ Facilitation is critical

  ▷ Requires back-up information

  ▷ Challenges design and operating decisions

  ▷ Identifies good practice

➤ Can be used at different project stages:

  ▷ Design

  ▷ Installation

  ▷ Testing

Figure 4-11 Example risk checklist

For each risk type typical risk scenarios were reviewed, enabling a quick 'score': total threats, total opportunities. The team were then able to develop an action plan to eliminate threats and maximize the identified opportunity:

➤ An example of a threat found was the potential for the product to be exposed to contamination from the facility environment.
➤ An example of an opportunity found was the potential for the facility to be reduced in overall floor area by better designing material and personnel flows, and reducing the size of areas which required a controlled environment.

Risk checklists have been used for many types of risk assessment and can be customized for a specific project or project area. For example, on a business change project the team identified that the people impacted by the business change represented the biggest risk to success of the project. They developed a series of checks which could be completed by observation and some targeted questioning and as a result they were able to trend a risk check score.

Tool: SWOT Table

The aim of this tool is to support Project Teams in structuring project SWOT analyses appropriately. Too often teams go as far as to identify their SWOTs but never actually do anything with the results. The

SWOT Table (Table 4-4) is a reminder that the ultimate aim of a SWOT analysis is to mitigate risks to the project through:

- Strengthening key strengths.
- Eliminating critical weaknesses.
- Maximizing the opportunities for success.
- Minimizing the threats to success.

Table 4-4 SWOT Table explained

Planning Toolkit – SWOT Table			
Project: *<insert project title>*		**Sponsor:** *<insert name>*	
Date: *<insert date>*		**Project Manager:** *<insert name>*	
SWOT identification and analysis			
Strengths	**Ranking**	**How strengthen?**	
<insert an identified project strength>	*<insert ranking>*	*<if this is ranked in the top 3 then insert the action plan to further strengthen this area>*	
<insert strength n>			
Weaknesses	**Ranking**	**How eliminate?**	
<insert an identified project weakness>	*<insert ranking>*	*<if this is ranked in the top 3 then insert the action plan to eliminate this weakness>*	
<insert weakness n>			
Opportunities	**Ranking**	**How maximize?**	
<insert an identified opportunity for the project>	*<insert ranking>*	*<if this is ranked in the top 3 then insert the action plan to maximize this opportunity>*	
<insert opportunity n>			
Threats	**Ranking**	**How minimize?**	
<insert an identified threat to the project>	*<insert ranking>*	*<if this is ranked in the top 3 then insert the action plan to minimize this threat>*	
<insert threat n>			
SWOT summary			
<insert a summary of the analysis and the ability of the project to resource action plans, thus reducing the risk profile for the project>			

SWOT identification and analysis

Brainstorm all possible SWOTs, review each category and rank 1st, 2nd and 3rd most important to the project. This can be done through team voting. Only rank the top three, as the maximum number of action plans that will need to be resourced and tracked will then be limited to 12. There needs to be some limit as projects do not have infinite resources available.

SWOT summary

The SWOT summary should provide conclusions on the risks identified, and the most important action plans which will mitigate the risks that need to be resourced.

As with any action plan, the SWOT actions need to be tracked during the remainder of the planning stage and into the delivery stage as applicable.

Tool: Critical Path of Risks Table

The aim of this tool is to support Project Teams in reviewing their critical path of risks. Too often teams get lost in the detail of risk assessment, generating long risk registers with no understanding of the cumulative impact of all risks and risk types. The Critical Path of Risks Table (Table 4-5) is a reminder that a Project Manager needs to continuously forecast the level of certainty of outcome for the overall project whether positive or negative.

Table 4-5 Critical Path of Risks Table explained

Planning Toolkit – Critical Path of Risks Table					
Project:	*<insert project title>*		**Sponsor:**	*<insert name>*	
Date:	*<insert date>*		**Project Manager:**	*<insert name>*	
Risk detail					
CSF	**Red risks**	**Amber risks**	**Green risks**	**Total risks**	**Probability of success**
<insert CSF 1>	*<insert number of red risks>*	*<insert number of amber risks>*	*<insert number of green risks>*	*<insert total number of risks>*	*<insert low, medium or high>*
<Insert CSF n>					
Risk summary					
Probability of success	*<insert low, medium or high>*	**Comments**	*<insert comments on the areas of risk focus for the project and the likelihood that the probability can be changed through risk mitigation>*		

Risk detail

For each CSF the risk distribution should be collated and analyzed. Typically, the following guidelines are used to assess the probability of success for a CSF:

- If there are any red risks in the risk profile, the probability of success is less than 50%: low. This is based on the assumption that there is at least one risk which has a high probability of occurrence and will prevent achievement of the CSF if it occurred.
- If there are no red risks and the proportion of amber to green risks is greater than 50%, the probability of success is 50%: medium. This is based on the assumption that the majority of risks to CSF achievement are amber.
- If there are no red risks and the proportion of amber to green risks is less than 50%, the probability of success is greater than 50%: high. This is based on the assumption that the majority of risks to CSF achievement are green.
- If there are no red or amber risks then the probability of success is high.

Risk summary

The risk summary is an opportunity to review the risk distribution across the path of CSFs. This is the critical path of risks. Typically, the following guidelines are used to assess the probability of overall project success:

- Any one red CSF and the probability of project success is less than 50%: low. This is based on the assumption that for a project to be successful all CSFs need to be achieved.
- Any one amber CSF and the probability of project success is 50%: medium.
- All green CSFs and the probability of project success is greater than 50%: high.

Tool: Risk Management Strategy Checklist

Inexperienced Project Managers often ignore the breadth of risk management strategy definition as they focus on identification rather than mitigation. The aim of the Risk Management Strategy Checklist (Table 4-6) is to remind Project Managers of the overall risk management process which needs to be followed during the development of the project control plan. The sections of the checklist follow the generic risk process as outlined in Figure 4-3.

Table 4-6 Risk Management Strategy Checklist explained

Planning Toolkit – Risk Management Strategy Checklist			
Project:	*<insert project title>*	**Project Manager:**	*<insert name>*
Date:	*<insert date>*	**Page:**	1 of 2

Risk identification strategy

What is the proposed set of risk methodologies to be used on this project?
<insert the list of risk methodologies and tools to be used on the project>
How will the methodologies be used to generate a full understanding of project risk?
<insert a brief purpose statement for each risk methodology identified. For example a SWOT analysis may be used for team building, an FMEA for design risk identification and a critical path of risks for an overall view>
How will risks be categorized?
<insert explanation of how overall risks will be categorized, via CSF, via consequence, via project stage>

Risk analysis strategy

For each methodology chosen, has a scoring system been defined for impact and probability?
<insert yes/no and attach the scoring system for each selected methodology>
Is it clear how risks will be detected?
<insert explanation of how the risks will be detected during project delivery>

Risk response strategy

Have all identified risks been prioritized?
<insert yes/no>
Have mitigation plans for all high-priority risks been developed?
<insert yes/no and attach mitigation plans>
Can all high-priority mitigation plans be resourced?
<insert yes/no>

Risk review strategy

Has a risk review strategy been defined for this project?
<insert yes/no and attach details of the method and timing of project risk reviews>
How often will the risk response plan be reviewed during the project?
<insert yes/no and any further actions required>

Risk summary

Summarize the status of the risk management strategy for this project
<insert any issues and the way that the project will have to deal with them>
Is the strategy suitably defined for the start of project delivery?
<insert yes/no and any further actions required>

Contract and supplier strategy

The majority of projects require some type of resource which is external to the organization, no matter the type of project. A control plan needs to understand the split of internal and external resources and ensure that this will deliver a successful project outcome. In its simplest terms there are four types of delivery resource (Figure 4-12):

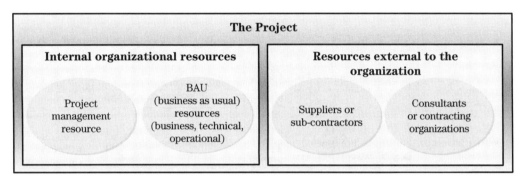

Figure 4-12 Example project resource review

> **Internal**
>> **Project management resources**: At some level most projects are assigned to an internal Project Manager, sometimes whether they have the capability or not. In addition the Project Manager may have other project-based resources available. For example, an organization may have a project management team available to deliver specific types of projects such as engineering or process improvement. Doing projects is their 'day job'.
>> **Business as usual (BAU) resources:** Depending on the type of project and organization there will be an assigned Project Team made up of internal resources from the business. These can be technical experts, end users or even resources doing their 'day job' in a project environment. For example, for a manufacturing facility project typical internal resources assigned to the Project Team would be those who are involved in the day-to-day operation of the facility: engineers, operators and quality/compliance resources.
> **External**
>> **Suppliers or sub-contractors:** These are resources which supply materials or equipment as well as the resources to design, install and sometimes test them. This is the most common external resource as most projects procure assets to deliver their project.
>> **Consultants or contracting organizations**: These can be complete organizations delivering a large part of the project scope or they can be specialist consultants with a specific technical or project management expertise.

Dependent on the project type, size or timing versus other business priorities the organization may choose any mix of the above. This is the first stage in developing an overall strategy for delivery of the project and is usually defined within the PDP. The control plan needs to define what external resources are needed and why, and then define how the resource is to be procured so that the project objectives are met and risk is appropriately managed.

The procurement lifecycle

Procurement is the usual name given to the process of engaging with resources external to an organization. Most organizations will have their own procurement procedures which a Project Manager will need to comply with, however within the PDP it would be usual to define how all stages of procurement are to be performed.

All resources external to an organization go through a similar procurement lifecycle (Figure 4-13). It does not matter if the resource is a service (person or organization) or a product (equipment, materials) the lifecycle remains applicable.

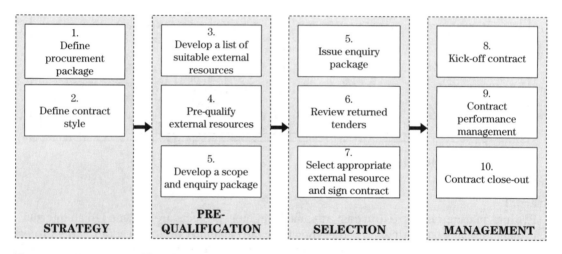

Figure 4-13 Procurement lifecycle

The four stages are discussed in more detail in the following sections.

Contract strategy

Define procurement package

Once a decision has been made to procure a part of the resource externally the Project Team need to divide the selected scope into various packages. This is the first level of contract strategy definition. Figure 4-14 shows a simple contract strategy for a small engineering project concerned with the installation of a new drier into an existing facility. This contract strategy defined:

- 11 procurement packages of which 5 required external resources.
- The boundaries of each procurement package, whether interfacing with internal or external sourced packages.

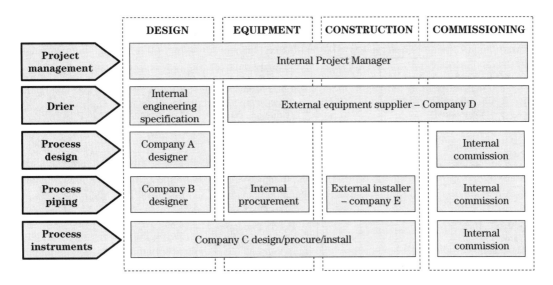

Figure 4-14 Example contract strategy

The selection of internal or external resource was based on level of expertise and whether it was available internally. The Project Manager also considered the risks in breaking the scope into either smaller or larger packages. This decision is usually based on how the performance tests of the final package of scope can be completed and who should be held accountable for that performance. However, other project drivers also impact the decision:

- **Cost:** It may be easier to gain larger discounts with larger packagers or it may cost less to go for smaller work packages that are managed in-house.
- **Schedule:** A larger package may take longer to pull together and then the supplier is completely in control of schedule, but equally smaller packages may have dependencies which cause delays if not managed appropriately.

For example, a business process improvement project had identified a scope consisting of:

- The development of a new business process (and associated SOP) to manage assets within a large organization having 5 separate facilities.
- The development of a new IT system which interacts with the business process in a lean way and which delivers accurate reports on current asset value and asset location.
- The design, installation and testing of the IT system and business process so that it is effectively integrated into BAU.

Two options for the contract strategy were developed (Figure 4-15).

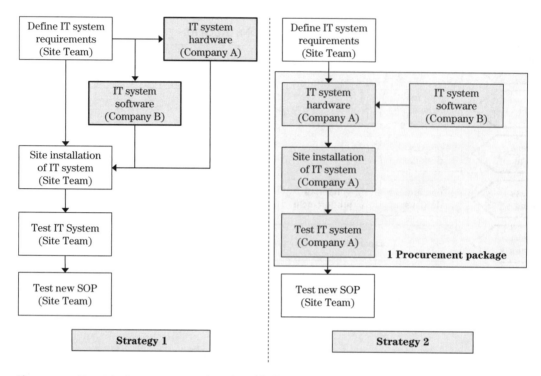

Figure 4-15 Example procurement package boundaries

Strategy 1 (Figure 4-15) was the usual way which business processes were improved: do as much as possible in-house and then buy the equipment or services needed as separate packages. This meant that the in-house Project Team had to piece together the final deliverable and were completely accountable for its performance. Although this gave some element of control to the Project Manager and also some cost benefits, the biggest risk was that the IT system wouldn't work as intended. In order to get either the software or hardware suppliers to take on a performance guarantee a larger procurement package had to be collated, as described by strategy 2 (Figure 4-15). This was eventually the selected strategy, with the organization having one contract with Company A. The benefits were:

- Using the external expertise in IT software and hardware to take accountability for the performance of the system in line with design needs. This expertise was not in-house. The system was handed over complete and tested and met all agreed performance acceptance criteria.
- Schedule risks were lowered as all interface management was with one company. Getting certainty from the small in-house installation team was an ongoing organizational issue and this risk was eliminated.
- The initial business process design had to be more robust than was usual for in-house projects to be an appropriate input to the contract. The contract strategy forced this robustness and the project benefited from it.

The initial costs estimated as a result of this contract strategy were higher than anticipated, but the usual 15% cost growth was limited to 3% mainly as a result of a clear fixed price contract. However, more critically for the organization, the business process was made 'live' in advance of business expectations largely due to the integration of internal business process design and external system design and delivery. The business benefits were therefore realized earlier than the business case demanded: asset management became a more accurate and efficient process.

As this example demonstrates, it is important to consider where responsibility for delivery of specific project requirements lies. This should be considered in line with the project goals and requires a review of the risk to the project and where the risks should lie. This is a balance; there may be increased cost in order to obtain a simpler solution for the Project Team, reducing overall project risk. It is important that the management of risk lies with the resource (internal or external) that actually has the responsibility and also the capability to manage it.

Define contract style

Once a decision has been made on the way that the scope is to be divided into in-house or external procurement packages, the contract style for all external contracts has to be defined. Although there are many different contract styles and types (dependent on the project type, the industry and the organization) these can be divided into two main generic types:

Fixed price

Where scope is fixed and design totally complete and cost certainty is important to the project then fixed price is an appropriate contract style. Because of the fixed scope and completed design there is likely to be very little change during the delivery of the contract and therefore very few, if any, contract variations. The contractor is free to work on the project with no discussion of the progress of his material or labour costs unless an interim milestone payment has been agreed within his fixed price. The risks associated with the fixed price contract are:

- Cost of change if you decide to change the scope.
- Potential for contractual issues requiring a higher degree of 'contractual' contract management.
- Unexpected schedule issues due to the lack of progress data (unless specifically requested as a part of the fixed price contract)

Reimbursable

Where design is incomplete cost is still important to the project, but delivery must commence in order to meet overall schedule requirements, then a reimbursable or schedule of rates type contract may be more applicable. As the design is only partially complete there are likely to be changes and ongoing design or delivery decisions. The use of schedule of rates or reimbursable type contracts makes all of these costs transparent for the Project Team, allowing them to make the most appropriate decisions for the project. There are many forms of reimbursable style contracts linked to the management and sharing of risk and rewards.

Contract style scenarios

Some typical project scenarios are presented along with the type of contact which has been chosen:

- *Design is only partially complete but final quantity of deliverables defined* ⇒ use a reimbursable fees contract with a guaranteed maximum fee based on the list of deliverables.
- *Time is the main project driver* ⇒ have target-reimbursable contracts with incentive clauses against key project milestones – completion of deliverables or commencement of beneficial use.

> *Clear complete scope with cost the main project driver* ⇒ fixed price contracts for large procurement packages with accountability for performance delivery (and therefore risk) with the suppliers.

For example, for the contract strategy outlined in Figure 4-14 for a small engineering project, the contract styles for the five external procurement packages have been defined (Figure 4-16). The organization looked at cost, schedule and scope drivers.

Selection of the correct contract strategy can increase the likelihood of a successful project outcome. This can be achieved by utilizing contract types which suit the situation and actively promote what is required for the particular project. Ultimately the selection of the correct contract strategy is about the management and control of project risk.

Supplier pre-qualification

Suitable external resources

Most organizations have processes to identify suitable external resources that they are happy to do business with. Normally an organization would expect to limit the total number of suppliers they use in order to best leverage existing and potential future relationships. Information from external sources is invaluable in getting another perspective on the suppliers under review:

> **Competitors:** An organization may have links with other similar organizations for the purposes of comparing information on supplier performance and capabilities.
> **Trade press:** This can give information on major contracts that the supplier may have recently won, new product developments or capabilities.

Company	Procurement package	Fixed fee (services) or fixed price (products)	Reimbursable fee (services) or schedule of rates (products)	Combined service/product supply fixed price
A	Process design package	Clear scope of supply ✔		
B	Piping design package	Clear scope of supply ✔		
C	Instrumentation supply		Using a preferred contract against a clear scope and schedule of rates ✔	
D	Drier package supply			Clear scope of supply and internal specification ✔
E	Piping installation sub-contract		Using a site contractor so using a schedule of rates for installation of piping procured internally ✔	

Figure 4-16 Example contract styles definition

- **Trade organizations:** These can be a useful source for potential suppliers particularly when sourcing products or services outside of an organizations normal operation. Registration with trade organizations who promote a quality standard can also be an indicator of a supplier's commitment to quality.
- **Quality systems:** Investigate the systems the supplier operates to and whether they align with those within your organization.
- **Company financial reports:** Financial information can be obtained from third-party organizations. The reports are based on statutory financial information which companies must provide. Their reports can provide a wealth of information about a supplier, names of directors, turnover and profit from the previous year which can be an indication of current or future profit levels/targets, it also provides information on current financial status and credit rating.

Pre-qualification

The aim of pre-qualification is to produce a list of external suppliers who are capable of meeting the requirements of the project, therefore supporting a successful outcome. This requires the consideration of many factors:

- **Capability:** Does the supplier have the correct capabilities?
  For example, if they are making a vessel, do they have experience with the materials, do they have equipment of the right size to produce the required vessel and can they produce the correct types of welding? In addition it is important to understand if their capability is completely in-house, and if not, where they will source it from.
- **Size:** Is the company a good size to work with the organization?
  For example, if they are either much smaller or much larger they may not be able to meet the requirements. If they are very much smaller they may not have all of the required capabilities, sufficient workforce to meet time demands, or sufficient quality control personnel. The order may also be too large for them to cope with financially, giving them far too much work in progress and straining cash flow. If they run out of funds part way through and cannot get materials supplied to them, then additional arrangements may be required to ensure a release of completed goods for the project. If they are much larger than your organization, then the order may not be a high priority as compared to work for their much larger clients.
- **Quality:** Can they achieve the correct quality standards?
  For example, do they have an appropriate quality system? Can they achieve the required levels of documentation and traceability? Do they have experience of working in the right industry sectors?
- **Ethics:** Do they operate the correct ethical standards? This consideration may be especially important when operating with in developing countries.
- **SHE:** Does the company have health, safety and environmental policies (SHE) and procedures which align to those in your own organizations?

Tender package development

A supplier can only be finally selected when they have been given the opportunity to respond to a formal tender package which is specific to the project requirements. There are three main components to a tender package (Figure 4-17):

- **Scope:** The documents that define what is required: the interim and final deliverables expected from the supplier as an integral part of the contract. A proposal from a supplier is only as good as the

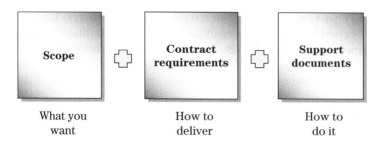

Figure 4-17 Tender package contents

scope they have been asked to quote for, therefore the development of a robust scope document for each tender package is required.

- **Contract requirements:** The documents that define the contractual relationship between the organization and the supplier including payment terms, confidentiality, intellectual property rights and also any schedule requirements. Supplier contractual responsibilities will usually be linked to information and/or decisions that the organization needs to provide to the supplier so that the scope can be completed. For example, a vessel manufacturer can contractually agree to a specific delivery date for vessel but this is dependent on Client approval within a specified time from request.
- **Support documents:** The documents defining the how the scope should be delivered, in terms of technical specification or format of the tangible output. For example, an organization may have specific quality standards they want the supplier to use.

Supplier selection

Issue enquiry package

Eventual selection is made much easier if there is a clear definition on how suppliers are to submit information. A Project Manager and Project Team need to be able to quickly identify if a bid is compliant or not. The project will have to compare the submissions from various suppliers, so it is imperative that they all provide the same information.

Apart from the information requested in the tender there will have been dialogue during the tendering process. In fair and objective selection processes any additional data generated during the tendering process should be sent to all suppliers so that each has the same information. This ensures that a like for like comparison can be made. Therefore, if one supplier raises a query then all suppliers should be advised of the requested information.

Bid appraisal

Bid appraisal is an important process and should be structured and objective. A part of contract planning is deciding how returned bids (sometimes referred to as proposals) will be analyzed. Typically factors such as cost, schedule, supplier proposed delivery strategy, proposed key resources and previous similar experience may be factored into the bid appraisal process.

However, it is important that a Project Manager never assumes that a supplier has quoted for exactly what was requested. A tender should be checked to ensure that the full scope of work has been quoted for. The like for like comparison is very useful if one supplier is much cheaper or much more expensive in any one area. In these cases, it is likely that they have made a mistake in understanding the scope or have purposefully excluded something. If schedule is important then ensure that similar assumptions on what a supplier may need from the organization are outlined. Typically a contract plan will identify the key criteria which are to be used to differentiate one supplier bid from another, the relative importance of one criterion over another and an objective scoring system (Figure 4-18).

	Requirement	Weighting	Required score	Vendor A	Vendor B
Criteria A					
Criteria B					
Criteria C					
Vendor approved?				✔	✘

Determine how scoring will be done – quantify soft issues   Apply to vendors and calculate score

Figure 4-18 Supplier bid review

For example, a product development project needed to find an external supplier to support the manufacture of a prototype medical device. Table 4-7 shows how two supplier bids were reviewed. Supplier B was selected based on the combined assessment of 'hard' and 'soft' assessment criteria:

➤ Supplier B did not have the lowest bid price but did have the lowest overall contract price when the cost risks were considered.
➤ Supplier B gained the minimum score in all other areas.
➤ Supplier B was able to provide the highest quality Project Team which had a major weighting in the bid appraisal.
➤ Supplier A gained a low score on overall cost due to its reputation for variation management and the view that additional variations would be seen due to the less robust scope.

Only 20% of the total score was actually related to cost. The majority of the selection was based on the certainty with which the supplier could deliver the required project outcome in terms of quality and functionality.

Table 4-7 Example supplier selection matrix

Criteria	Maximum score	Supplier A		Supplier B	
		Score	Comment	Score	Comment
Bid quality	5	5	Excellent	4	Some typos
Scope accuracy	15	14	One area has been misunderstood	15	Excellent
Delivery strategy	10	8	Manufacturing strategy needs discussion	7	Manufacturing strategy needs discussion
Bid cost	5	5	Lowest at $100,000	3	8% higher at $108,000
Anticipated total cost	15	8	Likely 20% variations based on reputation so would be over budget ($120,000)	14	Likely 5% variations based on reputation so would be within budget ($110,250)
Risk rating	15	10	Both schedule and cost risks	12	Some schedule risks
Reputation	5	3	Known for good quality but poor contract management	4	Known for good quality but with some schedule issues
Experience	5	5	Excellent	5	Excellent
Project Team	15	10	Weak in areas	13	Generally acceptable
Contractual	10	8	Issues with intellectual property clause	10	No issues
Total	**100**	**76**	**Not selected**	**87**	**Selected**

Supplier management planning

Contract launch

Once a supplier has been selected, final contract negotiations have to take place so that a contract can be signed. At this stage a Project Manager should have a clear idea on the contractual obligations he is entering, both in terms of what he expects to get from the supplier as well as what the supplier will need from the project to meet its obligations.

Within a contract plan it is usual to consider how to launch a contract. Typically a kick-off meeting will be held to confirm obligations on both sides of the contact. Both client and supplier will have obligations:

- Information.
- Design deliverables.
- Testing.
- Delivery.

These need to be defined and then tracked. If a client obligation is missed it will have an impact on the suppliers' ability to meet its obligations. For example, an equipment supplier requires specific data from the Project Team, and vice versa (Table 4-8).

Table 4-8 Example contractual obligations

Criteria	Client	Supplier
Information	▸ Provide decisions within an agreed time-scale ▸ Management of change within an agreed procedure ▸ Confirm operational philosophy	▸ Provide design information as agreed at kick-off (linked to client needs in other areas of the design) ▸ Review and assess impact of change requests within agreed time-scale
Design deliverables	▸ Provide user requirements specification (URS) as agreed ▸ Review supplier design documents within agreed time-scale	▸ Issue design deliverables on schedule ▸ Issue final documentation (including material certificates, etc.) at handover
Testing	▸ Provide performance criteria for testing with URS ▸ Be available for witnessing of testing	▸ Provide FAT (Factory Acceptance Test) and SAT (Site Acceptance Test) procedures ▸ Schedule FAT and SAT on schedule
Delivery	▸ Be available for receipt of goods on site	▸ Delivery to site on schedule

Contract performance management

Any external resource needs to be robustly managed in order to ensure that the required level of performance is achieved and the contract plan should indicate how this should be done. There are many methods of performance tracking:

▸ Progress can be checked against plan in order to assess percentage complete.
▸ The total number of tangible interim or final deliverables can be tracked. For example, the number of drawings produced or feet of piping installed can be compared against the total, in order to provide a percentage complete and an assessment of how much longer the task will take to complete.
▸ The criteria for measurement can be agreed with the supplier. This will allow both the supplier and the organization to check contract progress.

Audits can also be completed during the contract to assess various aspects and these should also be planned:

▸ *Quality* – it is important to ensure that the contract is being completed to the correct standard. It is better to check this along the way rather than waiting to the end of the project so that any problems can be corrected.
▸ *Safety and environmental audits* – checks should be made that suppliers are working to the correct safety and environmental standards as defined within their contractual arrangements.

Contract close-out and assessment

Contract close-outs reviews should be used to ensure successful completion of the contract for both parties. A contract close-out plan should include variation management, scope completion planning and feedback.

Contract variations

Contract variations require ongoing management as they have the potential to impact cost, scope and time project objectives. Therefore they should have been agreed and approved as they occurred, but in any event a contract close-out review should ensure that all such issues are finally agreed and closed. No more variations should follow. Contract variations are a normal part of cost control and need to be robustly managed within the boundaries of the requirements, contractual obligations, project budget and risk profile. They should be treated in the same way as any other type of change. A form of change control should be applied including logging of proposed changes, review of the proposal, approval or rejection and then the proposed change can be closed out. A contract variation may be simply that (a modification of a contractual term or condition) or could be related to a design change. Design change needs to be approved separately from the contract variation as the two processes are approving different things:

- A design change must match the scope needs of the project (quantity, quality, functionality).
- A contract variation is a discussion between two parties who have entered into a contractual arrangement to provide an agreed scope for an agreed level of compensation.

Contract variations which arise towards the end of contracts are more likely to be disputes in interpretation between the two contracted parties. Whereas those occurring during a contract are usually related to changes in scope or delivery strategy or the resolution of problems. A contract plan should make it clear how contract variations are to be managed (Figure 4-19) to avoid unexpected costs at the end of contracts which can then lead to contractual disputes.

Scope completion

Obvious though it sounds, there should be clear agreement that final deliverable(s) have been complete and handed over. This is a contractual requirement and usually allows final funds to be released to the supplier.

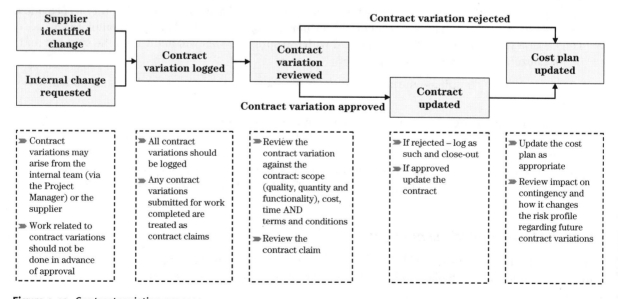

Figure 4-19 Contract variation process

Feedback

The most important aspect of the close-out process is open and honest feedback, especially where the two parties may work together in the future. Both parties have an opportunity to improve themselves. This opportunity should be fully realized and a typical process is an after action review (Melton, 2007).

Tool: Contract Plan

Often the contract strategy is left out of the PDP in the belief that there is time during early project delivery phases to find suitable external suppliers. The aim of the Contract Plan Tool (Table 4-9) is to support Project Managers in developing a contract strategy as a part of their overall development of the project control plan. It can also be used as a Contract Management Tool to review the progress of critical schedule milestones within the project.

Table 4-9 Contract Plan explained

Planning Toolkit – Contract Plan							
Project:	*<insert project title>*			**Sponsor:**		*<insert name>*	
Date:	*<insert date>*			**Project Manager:**		*<insert name>*	
Procurement package	**Contract style**	**Milestone dates**					
		Resource listing	**Tender package**	**Resource selected**	**Contract placed**	**Contract completed**	
<insert procurement package 1>	*<insert selected contract style>*	*<insert list of pre-qualified external resources>*	*<insert date>*	*<insert date>*	*<insert date contract to be placed>*	*<insert date>*	
			<insert tender package contents>	*<insert key selection criteria>*		*<insert acceptance criteria>*	
<insert procurement package 2>							
<insert procurement package n>							

Procurement package

Ensure that each procurement package is defined and listed. Check that there are no overlaps or gaps between packages (including scope being delivered by internal resources).

Contract style

The choice of contract style should be based on risk and align with project objectives of scope, cost and time. This may also direct which external companies are most appropriate to use. The mix of contract styles and interface between contracts needs to be reviewed for consistency and appropriateness. Bear in mind that the selected contract style will drive supplier behaviour and that suppliers working closely together on a project need to have aligning behaviours which match the needs of the project.

Resource listing

The list of pre-qualified resources (external suppliers) for each procurement package should be identified as a part of the plan. Often an organization will have a bounded list of external suppliers that they use for specific services or products. This is a sensible approach as it allows an organization to leverage cost and expertise across projects.

Tender package

To link in to the contract start and ultimate supply needs of the project, there needs to be a latest date for the collation and issue of the tender package to the selected resources. The contract plan should also clearly define the contents of each tender package to provide another link into the project scope definition.

Resource selection

There should be a standard and consistent time allowed for an external supplier to respond to a tender and then there should be a standard internal process to objectively evaluate all returned bids. The contract plan should highlight the key selection criteria for each procurement package. The contract plan should also define a latest date for selection of the supplier.

Contract placed

After selection there will inevitably be a process of final contract negotiation. If the scope associated with the tender package is unchanged, and the package is relatively simple, then this should occur soon after selection. However, for more complex packages the formal contract signing may take some time and this needs to be scheduled (time in the project schedule as well as project management resource to complete the negotiations). In terms of criticality this is often the one date which is on the project schedule as it represents the date when you expect work to commence on the scope area.

Contract completed

The end of a contract is just as important as the start but is often poorly managed. Apart from planning when a contract should be complete it is important to have acceptance criteria that define the agreed end-point for a contract.

How you plan to procure parts of your project will impact other parts of the project lifecycle. This tool challenges a Project Manager to plan the time-scales to get through the procurement lifecycle and to analyse how this fits with the overall project time-scale/strategy. Some contracts will inevitably:

➤ Take longer to kick-off due to the nature of the tendering process.
➤ Introduce cost risk early or even late in the project.

▶ Push risk one way (at a price) or the other.

▶ Support or limit 'fast-track' delivery.

The contract plan defines how the procurement packages have been defined and supports assurance that there is no duplication or gaps and that the various interfaces have been managed. The contract style used may depend on which resource is actually doing the procurement (internal or external). A contract plan is a baseline tool from which a Project Manager can track project procurement activities. It is useful to identify if any of the milestones are on the critical path and if so highlight these as critical milestones. Non-critical items are sometimes given 'start as late as' options and if not tracked can be easily pushed onto the critical path.

Tool: Supplier Selection Matrix

The aim of this tool (Table 4-10) is to support the articulation of project capability needs through the definition of key capability criteria and then the use of those criteria to select appropriate external resources.

Table 4-10 Supplier Selection Matrix explained

Planning Toolkit – Supplier Selection Matrix							
Project:	*<insert project title>*			**Sponsor:**		*<insert name>*	
Date:	*<insert date>*			**Project Manager:**		*<insert name>*	
Potential supplier	**Selection (yes/no)**	**Selection criteria**					
		Criteria 1	**Criteria 2**	**Criteria 3**	**Criteria 4**	**Criteria 5**	
<insert name of supplier under consideration>	*<insert decision>*	*<insert score vs criteria>*					
<insert name of supplier under consideration>	*<insert decision>*	*<insert score vs criteria>*					
Summary							
<insert any summary comments regarding the final selection and also any gaps which still remain, how these may be eliminated and any risks which need to be managed within the team>							

Potential supplier

At some stage in the selection process there will be a list of potential suppliers who need to be formally assessed. These companies may well have already gone through a first filter depending on whether the selection is for a specific contract or a general capability review (pre-qualification for use within a company).

Selection

A clear decision needs to be recorded. Any explanation of the decision can be put into the summary section if necessary. This is often needed if senior stakeholders have been championing a specific supplier however it is also good practice to track selection decisions for use in future projects within the organization.

Selection criteria

The criteria should be based on some element of capability: knowledge, skills, behaviours and/or experiences (proven performance). They should be able to be used to objectively assess a supplier and be capable of allowing comparison between suppliers. For example, selection of an engineering contractor may use 'experience in pharmaceutical sector' as a filtering criteria and then a more specific selection criteria 'have experience of current compliance legislation for medical devices' to differentiate between suppliers. In this way the length and type of experience can compared. Other typical selection criteria are:

- Total contract cost.
- Delivery date.
- Supplier team members (capability and experience).

For each criteria there should be a defined scoring system so that the most important criteria are given a higher weighting. In this way contract cost becomes one part of the selection process rather than being the main criteria.

Summary

The Project Manager would use this section to highlight gaps and current risks and how these are to be managed so that the project has the highest potential for success. Even within this type of procedure organizations expect some justification if the lowest cost is not selected. The selection criteria should consider total overall cost to the organization both in terms of 'hard' money which the supplier is paid (contract cost plus agreed variations) and the 'soft' money or 'value' which the supplier gives the organization. Such value might be:

- The early delivery of an item.
- The reliability of an item during its lifetime (and reduced lifetime costs).
- Exceptional project management requiring reduced contract management time.
- Increased quality or functionality of an item.

If the supplier is delivering additional value then the total overall 'cost' should take this into account and therefore be included in the supplier selection process.

Project control strategy

The majority of those who have previously been involved in a project will recognize the basic project control principles:

- Develop a cost and schedule based on the delivery of the defined scope.
- Develop a schedule which is logic linked and resource loaded.
- Use change management in conjunction with cost and schedule contingency management to maintain control.
- Align baseline plans to progress measurement and risk management so that forecasting is robust.

However, planning a control strategy is more than pulling together a set of tools. It involves the development of the philosophy and methodologies that the project will use to manage cost, scope and time so that the certainty of outcome is increased. To be in control of cost, scope and time a Project Manager needs to be able to forecast accurately. Control is not just about monitoring. A typical way that a control strategy is developed is shown by the activity tree in Figure 4-20. This diagram also highlights the key deliverables typically generated as a result of control planning. Additional tools are used to combine all three aspects to evaluate progress.

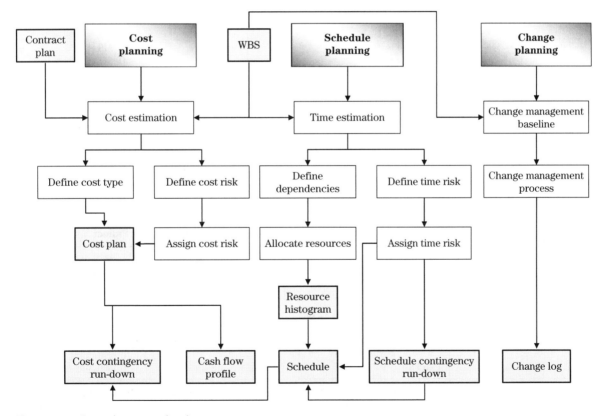

Figure 4-20 Control strategy development process

Cost planning

Most projects involve some form of cost planning although this may take different forms depending on the project funding strategy and the project type:

➤ Capital engineering projects usually require capital funds (for equipment and services) estimated to within ±10%.

➤ Business improvement projects usually require funds from operational expense budgets (for external consultants and/or materials) and internal resources and need to be planned at the start of each year.

The one thing that all projects will have in common is a cost plan (Table 4-11). Although there are a variety of formats a cost plan should define:

➤ What items are included in the cost plan (with links to the previously defined WBS and procurement package breakdown).

Table 4-11 Example cost plan

Item	Estimated cost	Estimation method	Estimation accuracy	Cost risk allocation	Cash flow profile
IT system	$65,000	Supplier fixed price bid for both scope items in one procurement package	±5%	Allow 5% as contingency = $3,850	M1 = 0 M3 = 50% M6 = 95% M7 = 100%
IT system installation	$12000				
Contingency budget	$3,850	Total based on cost risk assessment per item	n/a	$3850	Likely use half at M3 after design fixed and the remainder at the end of the installation in M7
Development budget	$3,850	Level of development work anticipated based on previous similar projects	n/a	Allow 5% of the external system capital costs	Likely use before M3
Total capital	**$84,700**	**Development and contingency budgets are owned by the Project Manager**			
Project management man-hours	140 days × day rate	50% time on project based on previous projects	±5%	Allow 5% in contingency	M1 to M14 equal amounts
Design man-hours	200 days × day rate	2 people 50% over design and then testing	±10%	Allow 5% contingency and 5% development	M1 to M6 then M10 to M14 equal amounts
Materials	$5,000	Purchase price	Fixed	None	M14 payment
Total operating expenses	**$5,000**	**For purposes of cost centre allocation only materials are considered cost in terms of financial governance. Internal resource costs are not specifically allocated to the project budget.**			

- The cost estimate for each procurement package or item in the WBS.
- To what accuracy they have been estimated and how.
- Specific cost risks associated with each item and therefore any contingency, escalation or development allowances.
- The schedule for cash flow (showing when funds will be spent).

As a result of the development of a cost plan the baseline cost report can be generated for use during project delivery. An example of a cost plan template is contained in the case study at the end of this chapter (page 218).

Within the planning phase it is important that a baseline is generated, although for most capital projects this will have been a requirement for project approval prior to detailed planning.

Estimation method and accuracy

Whether the costs are capital or expense, materials, services or internal resources, there should be a record of how the cost was estimated. This sets the parameters from which the Project Manager will control costs in the future. Typical methods of estimation include:

- *Historical data* – for example, basing internal man-hour totals on a similar previous project.
- *Supplier budget quotations* – this is where a draft specification or scope of services may be sent to a supplier for an initial cost estimate prior to the formal tendering process.
- *Factorial estimates* – this is where an organization has a database of previous projects and can apply some factor based on difference in scale. For example a new building may be estimated by considering its floor area and comparing it to a standard cost per floor area for a similar facility.
- *Schedule of rates* – this is where a supplier has a set rate for a specific quantity of supply. For example, a consultant would have a set day rate or a civil contractor may have a detailed schedule of costs for different types of materials supplied.

Obviously where external suppliers are providing the bulk of the project resource, the most accurate estimate will be when they provide formal bids prior to contract placement. This is unlikely to occur during the planning phase and so some cost risk needs to be taken when setting the cost baseline.

Cost risk allocation

The level of cost risk due to estimation accuracy is dependent upon when in the project roadmap the cost baseline was completed. For example:

- Business case development – confirming the concepts pre-approval typically ±30–50% cost accuracy.
- Business case approval – gaining funding approval against a clear benefits case typically ±10–15% cost accuracy.
- Once all procurement packages have been contracted typically ±5–10% cost accuracy.
- Once all contracts are complete and are being closed out typically ±1% cost accuracy or less.

However, cost risk from estimate inaccuracy is only one type of cost risk and others need to be evaluated (Table 4-12).

Table 4-12 Example cost risk assessment

Cost risk factor	Typical risk scenario	Typical risk response
1. Estimation accuracy	The cost estimation method coupled with status of scope definition introduces potential uncertainties: ➤ An equipment cost estimate at ±20% accuracy has a high potential of variation from the estimate	Some parts of the cost risk can be mitigated through having a development budget to allow for cost increases as scope accuracy improves, in other cases a contingency allowance is allocated to cope with more general estimation uncertainties. Risk cost assigned here is expected to be needed by the project although it is usual for a Project Manager to define the nature of the estimation inaccuracies.
2. Unit cost fluctuations	Some materials or services can have seasonal price fluctuations or changes in cost linked to other external issues: ➤ Currency ➤ Availability ➤ Current economic situation	Those parts of a cost plan which may be impacted are usually assigned an escalation allowance. Risk cost is assigned to a specific item and only used if the escalation scenario occurs.
3. Incomplete scope definition	Early estimates of a project cost may be based on conceptual scope or on a scope that has known gaps.	Allocate a project development allowance that is expected to be spent.
4. Contract style	A fixed price contract is more likely to have contract variations at the end of its life; a reimbursable contract has a different type of cost risk which is seen throughout the life.	Allocation of a cost contingency based on contract style. A fixed price contract would typically have a lower contract cost at the start of the project with a higher potential for cost variations towards the end of the contract. Cost contingency should be allocated to cover this scenario.
5. Supplier reputation	Some suppliers have a reputation for excellent services and poor project or cost management.	Ensure that the bid appraisal compares the lower cost of this bid with the anticipated growth cost risk.
6. Unexpected contract situations	Some areas of scope are more complex than others and the potential for unexpected issues is higher: ➤ Tests that have to be done twice ➤ A technical solution that doesn't work and needs to be changed	Allocation of a contingency based on the level of rework which may be needed.

These cost risk factors lead to two types of cost risk funding:

➤ *Escalation* – funds set aside to cover general market cost increases (page 112). These are not available to use for other areas of cost risk.
➤ *Contingency* – funds expected to be spent to mitigate cost risks. Some organizations separate scope development costs from general contingency but here contingency refers to all costs risks but escalation.

Whatever the definition of the various amounts of allocated cost risk, it is usual that this is managed separately from the individual procurement package budgets by the Project Manager. Typically contingency cash flow is planned and then tracked (Figure 4-21).

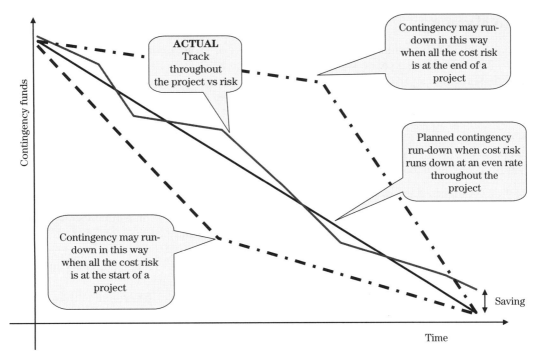

Figure 4-21 Cost contingency run-down

Tracking the use of cost risk funds is an important part of a control plan as it allows a Project Manager to gauge where cost uncertainty is in the project and when this uncertainty has passed. The case study at the end of this chapter (page 217) shows how a contingency fund has been sub-divided to take account of the two main cost risks. This sub-division is typically used to take account of the two main reasons for contingency and the different ways these are managed:

➤ Development funds are usually within the total contingency budget but assigned to a specific contract. The Project Manager would usually have a clear list of scope development areas with assigned development cost estimates.
➤ True contingency funds are usually not visible to external suppliers and may not be assigned to any one contract even though they cover a very specific risk scenario. They cover all areas of cost uncertainty as described in Table 4-12.

Cash flow profile

Once the cost plan baseline is clear the next stage of cost planning involves a review versus the schedule baseline. The cash flow profile (usually an 'S' curve) should be generated to give:

➤ The Project Manager an overall plan from which to assess project progress (page 199).
➤ The finance department a clear definition of when money will actually be leaving the company accounts.

Schedule planning

Of all the elements of project management (and within that of project control) that have made significant advances in recent years, schedule development has to be at the top. There are various terms which describe how fast a project can be done, from fast track, to lean, to agile. However what they all have in common is a desire to achieve greater project delivery speed whilst still meeting all required project success criteria. All schedule development methodologies have the following in common:

- They are based around a WBS.
- They take account of dependencies between activities.
- They look at resource loading.

Figure 4-22 summarizes the differences in three main types of schedule development methodology:

- *Traditional* – activities are linked via a robust critical path and duration estimates are based on previous projects. Schedule development is very risk adverse and projects tend to be based around standard WBS for a specific type.
- *Fast track* – activities are progressed 'at risk' as the critical path is challenged. This can lead to spectacular successes and failures as risk taking is successful or not.
- *Agile* – activities are reviewed for their value to the project goal and the critical path is challenged by reviewing dependencies and durations. A new schedule development culture integrates a 'leaner' approach and projects are customised to deliver specific value to a business.

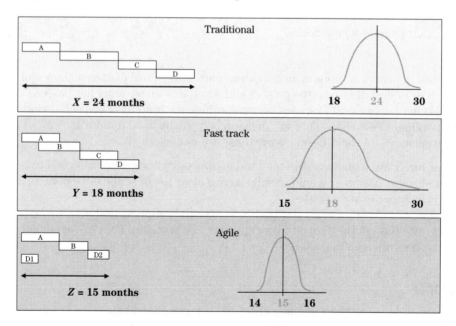

Figure 4-22 Comparing schedule development methodology

There are three concepts which work together to reduce project cycle times and produce an 'agile' result:

- Theory of constraints (TOC) (Goldratt, 1997).
- Lean six sigma (Adams et al., 2004).
- Schedule risk assessment.

Together they require a change to schedule development culture and also the way the schedule is managed during delivery.

Schedule development process

The basic process to develop a schedule remains valid:

- Use the WBS as the basis for the schedule. If the schedule is being reviewed in advance of robust WBS development then the team should brainstorm the activities required to deliver the final deliverable first.
- Review any dependencies between the activities in the WBS (Figure 4-23) and confirm if these are time, resource or outcome dependencies.
  - Time dependencies are activities that will take a set time no matter the resources used or require a specific lag between activities which again is not resource dependent.
  - Resource dependencies are activities which use a specific resource and therefore availability of this resource can change the time that should be allowed in the schedule. The resource could be an asset or a person.
  - Outcome dependencies are activities which rely of the progress of another activity and are the dependencies which are commonly used for preliminary schedules when resources have yet to be considered (Figure 4-23).

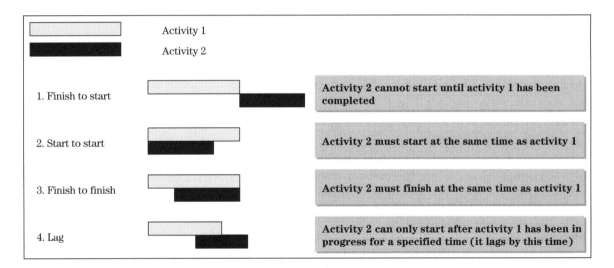

Figure 4-23 Activity dependencies

- Once all dependencies have been identified a critical path should be easily identified (Figure 4-24) – this is the first critical path analysis (CPA):
  - The critical path is the linkage of activities which have no 'float' from the start to the end of the project.
  - It describes the shortest route through the project and typically determines the project end date goal.
  - Other activities, not on the critical path, have 'float' allowing them to be started later than scheduled without impacting the project end date.
- Once the initial CPA has been completed it is usual to review each activity in more detail and to test duration and resource assumptions.

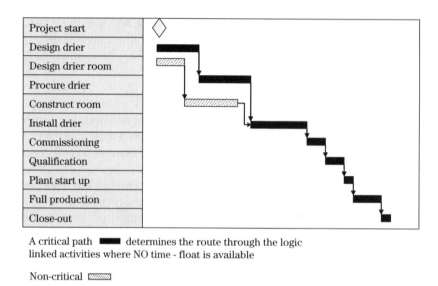

A critical path ▬▬▬ determines the route through the logic
linked activities where NO time - float is available

Non-critical ▨▨▨

Figure 4-24 Critical path analysis (CPA)

- The CPA is then reviewed with all changes in place and the project end date confirmed.
- At this stage a schedule is usually 'baselined'. This effectively freezes the activity start and end dates and is usually used to produce a milestone progress chart for the Project Team.
- During delivery the baselined schedule is used to track progress, with any deviations being noted versus the critical path and associated end date.

Schedule development culture

There are many reasons why there is a continuing focus on project speed, but generically most projects need to be complete as soon as possible to start to realize the benefits from the investment in the project – whether this is capital or expense investment. Much has been done over the years to speed up projects but many of these have introduced further risks themselves. This is because they have not understood the true cause–effect relationships within schedule development and the culture which drives schedule development behaviours.

If we look at an example pharmaceutical facility project delivered 10 years ago (approximately £15 million, greenfield site) and then an equivalent one delivered 5 years ago, we would see a positive trend in terms of project delivery speed (Table 4-13).

Table 4-13 Trends in project delivery speed

Schedule KPI	Project A 10 years ago	Project B 5 years ago	Project C vision
Average project cycle time (months)	24	18	15
Schedule variance – late (months)	6	12	1
Schedule variance – early (months)	6	3	1

- *Project A* – describes a very traditional approach to schedule development based on previous norms both in terms of activity duration and dependencies. Both include a high degree of 'safety' to cover both real and perceived risks. Variation over a number of projects is normally distributed and project success is predictable to within 6 months.
- *Project B* – describes a fast-track approach where activity durations have been left as per the norms but the dependencies have been challenged and/or ignored. As a result increased risks are being taken and whilst overall project cycle time reduced, there is an increased potential of a higher schedule variance (late). As a result overall predictability of project success decreases.
- *Project C* – is the vision for future projects. The decreased variation is the vision of predictability and the decreased cycle time is based on a different culture of schedule development and management of delivery.

If we root cause 'why schedules are not met' then we come up with some interesting results. The majority have a link to both schedule development as well as schedule management during delivery.

Schedule development

- Activity duration estimation activities are flawed and/or based on inappropriate norms and integrate a safety factor within them.
  - We base our activity duration estimates on 'norms' rather than reality.
  - We ask our team for certainty so they add safety into their estimates.
- We build a critical path based on our previous understanding of activity dependencies without estimating the resources we intend to use.
- We don't assess specific schedule risks during estimation.
- We plan to start 'as early as possible' to keep activities off the critical path.
- We develop a list of milestone dates.

Schedule management

- Unexpected things happen for which we have no contingent response.
- Expected things happen for which we have no contingent response.
- When we do things quicker than estimated we can't use the time because resources are not available (due to the focus on due dates).
- Schedule contingency is 'lost' within each activity and then used inappropriately.
- The team focus on meeting milestone dates (critical and non-critical) not the ultimate project deadline – we get caught up in the tasks at the expense of the project goal.
- Activities always start as late as possible so any integrated safety not required can't be used.
- Resources are often distracted by other work outside of the project – multitasking.

Advances in schedule development

There seems to be a myth that the critical path of a project is fixed for a specific type of project and that to speed things up we have to introduce and then mitigate risks by ignoring dependencies. For example, when we send out tenders for a package of equipment before the detailed design is complete we are risking potential rework if the design basis changes; the benefit of purchasing a long lead critical path item as early as possible appears to outweigh these risks.

Traditional schedule development processes are limiting the reduction in project delivery time-scales and therefore it is appropriate to consider three concepts which are changing the culture of schedule development and management:

➤ The TOC.
➤ Lean six sigma.
➤ Schedule risk assessment.

Theory of constraints (TOC)

The TOC has been successfully applied to project management and is certainly one of the key concepts in changing schedule development culture (Goldratt, 1997).

➤ TOC focuses in on the bottleneck within a process and then works to 'exploit' it, subordinating all other processes to it as you can never move any faster than the bottleneck (any resource whose capacity is less than the demand placed upon it).
➤ It looks at dependent events and their statistical fluctuations.

When a project schedule is developed, a list of dependent activities (WBS) is defined along with an estimate of their duration (thus introducing statistical variations). In doing so a critical path is built where the bottleneck is often resources required to perform critical activities. During this process it is important to understand:

➤ Why things are dependent: duration or resource.
➤ The impact of variation: what's the likelihood that a chain of activities of average capacity 'x' each will actually deliver 'x'.

TOC proposes the development of a critical chain as opposed to a critical path. If resources are always available in unlimited qualities then the critical chain is the same as the critical path. It proposes that:

➤ The schedule is built initially without any 'safety' – recognizing that there may be only a 50% confidence of achieving these.
➤ Resources are 'ring fenced' and categorized as those doing critical or non-critical tasks.
➤ The schedule is *not* managed by 'task due dates' but by 'time to complete'.
➤ Schedule buffers are defined based on identification of likely task variation and placed within the schedule to protect critical activities from non-critical ones.

In this way a project timeline can be significantly reduced.

Lean six sigma

A complementary principle is lean six sigma (Adams et al., 2004) which focuses on:

➤ The identification of value.
➤ The elimination of waste.
➤ The reduction in variation.
➤ The release of value flowing to the customer.

In terms of applying this to schedule planning the following activities need to be performed:

➤ Challenge the traditional WBS versus the definition of customer value.
➤ Eliminate all non-value adding activities – these are waste. For example a design deliverable which is never used to procure or construct something.

Allow value (activities, information and decisions) to flow. For example, identify ways of working (WoW) which cause issues to the flow of activities, information and decisions such as functional behaviours and adversarial contractual relationships.

Lean six sigma tools are challenging Project Managers and Project Teams to completely understand what value means to their project and the most appropriate way to deliver it.

Schedule risk assessment

Risk assessment is another part of the 'new culture' for schedule development and management. Whereas previously all risks were treated similarly (in a risk adverse culture) or conversely ignored (in a risk taking culture); now risks are identified, assessed and an appropriate risk response defined based on risk impact. Schedule risk assessment is one specific tool used to allocate schedule risk on an activity-by-activity basis related to the potential impact of not meeting the schedule. It also proposes the collation of schedule risk into buffers which can be run-down as the project progresses, recognizing when risks have not occurred and time has been saved. It is analogous to a cost contingency run-down methodology. Schedule risk assessment enables more effective and consistent risk-based decisions.

This concept works with TOC and lean six sigma as it challenges the WBS (does each activity add value?) and asks the team to trust the Project Manager (who ensures that performance measurement is not against individually assigned activities with removed 'safety' but aligns with the overall project outcome and shared team goals).

A new schedule development culture

The three concepts introduced lead to a schedule development and delivery culture which can reduce overall project cycle times predictably. However these concepts require a culture change where:

- The quality of schedule estimation is increased.
- Resources and resource dependency are considered more overtly with critical resources 'ring fenced' for the project.
- All schedule risks are collated into various schedule buffers and placed where they are most likely to protect the value delivery to customer (Figure 4-25).

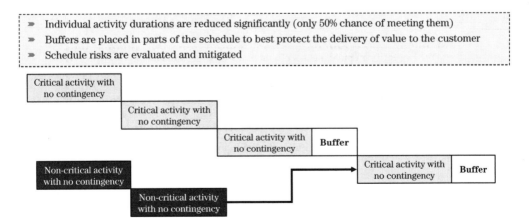

Figure 4-25 A new schedule development culture

- Activities are evaluated for 'value-add' and those that are waste are removed.
- Critical dependencies are managed rather than 'management by milestone'.

This level of change is not easy and it has taken time for some of these concepts to be tested, however benefits claimed by these concepts are noted:

- The TOC methodology (critical chain) reports a 95% on-time and on-budget completion versus benchmark data for comparable projects which are generally delivered late.
- The application of lean six sigma has decreased project cycle times and overall project costs of the order of 15% in early projects, with the potential for further decreases.
- The use of 'best practice' risk management techniques has enabled both of the above to be effectively applied.

The business (the customer) has an increased assurance that the project will be delivered when stated. For example, a facility shut-down project has completed a schedule risk assessment and identified schedule contingency for all activities. These were placed in a schedule buffer immediately before start-up to protect this critical milestone (Figure 4-26).

Scope area	Estimate	Allow	Dependency	Into buffer	Comment
Reactor module					
Design	4–6 Weeks	4 Weeks	**MEDIUM** Linked to integrated engineering design	1 Week (pre-shutdown buffer)	Need to consider resource levels
Procurement	12 Weeks	12–16 Weeks	**HIGH** Needs completed design	2 Weeks (pre-shutdown buffer)	CRITICAL PATH
Installation	4–6 Days	4 Days	**MEDIUM** Needs Civil Works completing	1 Day (shutdown buffer)	Day 3 of 15-day shutdown
Testing	5–11 Days	5 Days	**HIGH** Needs complete system installed	3 Days (shutdown buffer)	

- The Project Team build the work breakdown structure from an understanding of value specific to the customer of their project
- A schedule risk assessment is completed and contingency allocated to project buffers. (one prior to the start of the shutdown and one prior to the end)
- Activity durations are linked to specific resource profiles

Figure 4-26 Example schedule buffer development

In another example the application of lean six sigma and schedule buffer placement is shown as a 'before' and 'after' exercise (Figure 4-27).

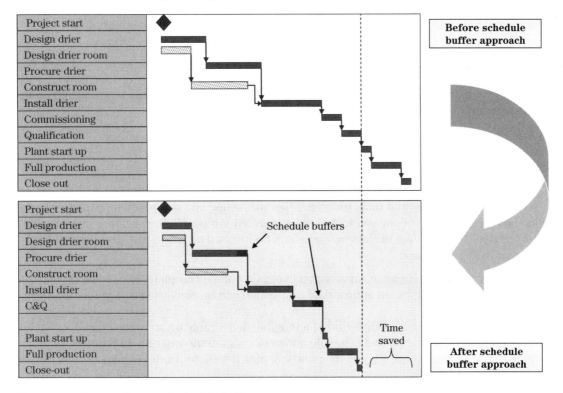

Figure 4-27 Example impact of schedule buffer use

As a project is delivered some activities will be completed within the estimated time and others will not, requiring the expenditure of the schedule buffer (Figure 4-28).

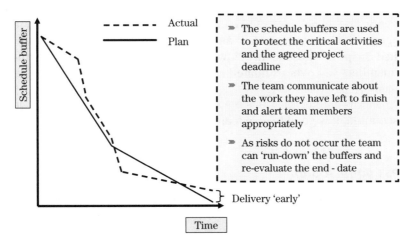

Figure 4-28 Schedule buffer run-down chart

If schedule development culture can adapt to the principle of schedule buffers, schedule risk assessment and removal of wasteful activities then it is clear that projects could be delivered quicker. However, it is important that the Project Team trust that they will not be measured against their 'best estimate' in terms of performance and they need to communicate continuously on how activities are progressing.

Progress measurement

A crucial element of any control plan is the definition of how progress will be measured and analyzed. We need to measure progress to:

➡ Confirm project status.
➡ Assess performance against the plans: 'are we going off track?'
➡ Support forecasting (cost and time to completion).
➡ Enable progress reporting.

The chosen methodologies and tools for cost, schedule, scope and risk planning will already determine how progress can be measured and then analyzed. Above all, the progress metrics should support delivery of the specified project outcome and be focused on the key project drivers. Typical progress measures systems are:

➡ Team member opinion – subjective assessments of percentage complete.
➡ Progress against plan – using an appropriate unit which can be converted to a percentage complete or a cost ratio.
➡ Achievement of milestones – completion of a tangible deliverable on an agreed date.
➡ Earned value – reviewing cost and schedule adherence simultaneously by tracking the actual completion status of deliverables and the cost/schedule it took to produce them.

Team member opinion

Team leaders are expected to manage and control activities within their area, so one measure of progress is their view on what activities have been performed and to what level of completion. Although this can be a useful guidance it is unlikely to surface any problems until an activity is nearing completion. For example, a mechanical design team leader was reviewing the progress of design drawings for each area of a new facility. Each week he would collate the percentage progress, only reviewing actual deliverables if the progress didn't match the plan. As the critical milestone for handover of the deliverables approached the progress suddenly slowed until it remained at 95% for a number of weeks. In this case the team members had over estimated early progress and the team really needed to approach progress in a more objective way.

Within the control plan it is unlikely that this method on its own is robust enough, although it should not be completely discounted. It is good to give team members responsibility for measuring their own work progress.

Percentage progress against plan

The majority of the planning tools will allow a simple measure of progress to be used for a quick overview. For example, the cash flow analysis in Figure 4-29 is useful in raising issues:

➡ If contracts are being placed and committing more or less cost than planned (a possible early indicator that bids are coming in higher or lower then estimated).

- If money is being spent too slowly or too quickly (a possible early indicator that some contacts are invoicing more or less than anticipated by the contract plan).
- If contingency is being used earlier in the project than anticipated (a possible indicator of estimate inaccuracies, contract mismanagement or major technical issues).
- A cost ratio can be calculated from the amount of money spent (usually invoiced rather than committed) to the amount of money that should have been spent (as planned via the cash flow profile).

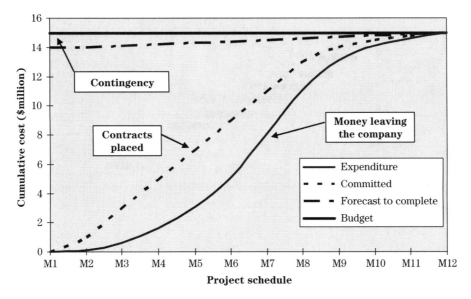

Figure 4-29 Example cash flow progress measurement

In terms of fixed price contracts a percentage completion is usually the only indicator of progress until deliverables are due. It is therefore important that the contract is set up so that progress can be planned. For example, if a fixed price contract to develop some software is due to take 12 weeks there would be an expectation that at 6 weeks the progress reported would be 50%. However unless this is aligned with some expectation of interim deliverables the Project Manager has no way to ratify this progress. When measuring progress within the Project Team it is usual to develop an objective system which tracks deliverable development. For example, a design deliverable may have four distinct states from start to completion:

- No work started = 0%.
- Draft design deliverable = 35%.
- Deliverable reviewed by peers = 70%.
- Deliverable approved = 95%.
- Deliverable issued to supplier = 100%.

In this way the total percentage progress from all design deliverables can be collated and compared with the planned total percentage completion.

Milestone progress

As explained above, it would be usual in most control plans for a Project Manager to have some progress measurement system which is linked to the delivery of tangible interim deliverables. This can be done in a number of ways.

A tracking Gantt chart (Figure 4-30) is perhaps the most common method of assessing the achievement of progress. It relies on the proven delivery of tangibles against plan. In some regards this method can work on subjective data but for an overall project picture this is not advised.

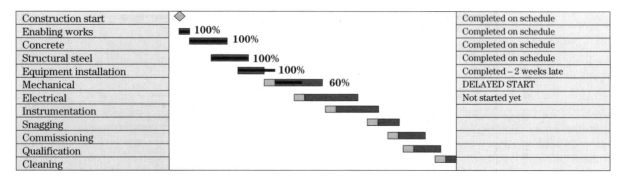

Construction start			Completed on schedule
Enabling works	100%		Completed on schedule
Concrete		100%	Completed on schedule
Structural steel		100%	Completed on schedule
Equipment installation		100%	Completed – 2 weeks late
Mechanical		60%	DELAYED START
Electrical			Not started yet
Instrumentation			
Snagging			
Commissioning			
Qualification			
Cleaning			

Figure 4-30 Example tracking Gantt chart

A milestone chart is often generated from a Gantt chart, highlighting the interim deliverables required at specific times. It would be usual to only identify critical milestones.

Deliverables matrices can also be used to track milestone progress against deliverable status (Table 4-14). This system looks at the value of work performed by linking progress to deliverable status (as described earlier) but then aligns this to a critical milestone date. For example, in Table 4-14, the status of a pipe work installation is tracked. For every system of pipe work five milestones are highlighted, each having a significance to other elements of the project and are critical in terms of their dependencies. In this example the payment terms agreed with the installation sub contractor were also linked to this progress measurement system.

Table 4-14 Example deliverables progress system

Deliverable status (for each of 20 systems)	Complete (%)	Milestone date
Isometric complete	10	Week 4
Pipe work has arrived on site	30	Week 6
Pipe work installed	75	Week 12
Pipe work tested	90	Week 15
System handed over	100	Week 16

Earned value

A Project Manager ultimately wants to track the progress of cost, scope and schedule objectives and an 'earned value' methodology builds on previous progress measurement systems as it aims to:

- Establish the value of work actually done.
- Compare this against the elapsed time spent doing the work.
- Compare this against the resource or cost spent doing the work.
- Interpret the above to forecast the cost and time to completion.

Earned value requires:

- A plan linking actual deliverables to cost and time.
- A method to objectively assess the progress of the deliverable, for example, metres of pipe work, volume of concrete, number of specifications, etc.
- Inspection of the work to check progress.

Earned value can be used during any stage of project delivery and for all types of projects. Traditionally earned value has been strongly linked to construction projects and whilst the level of tangibles on a construction site make this an appropriate tool to use, all projects have tangible deliverables that can be measured by earned value. For example, within any project with a design phase, the design deliverables can be used:

- Number of specifications completed.
- Number of drawings developed, checked and approved.

In addition, all projects which use external resources can use procurement tangibles:

- Number of tender packages issued (or financial value).
- Number of purchase orders completed (or financial value).
- Number of supplier documents received and reviewed.

Whilst in construction and testing the following deliverables are used:

- Amount of concrete poured.
- Number of floors or walls erected.
- Meters of pipe work installed.
- Meters of cable installed.
- Number of systems tested.
- Number of systems handed over.

Within business change projects the tangible deliverables typically used are:

- Number of workshops held.
- Number of business processes developed.
- Number of root cause issues dealt with.

These lists are not exhaustive and all can be used to give an accurate picture of overall project status in terms of cost and schedule adherence and therefore forecast to completion.

Earned value uses cumulative cost or man-hour 'S' curves as the basis. This format of 'S' curve is the graphical representation of the cumulative expenditure of a set amount of funds over a planned time-scale. Three 'S' curves are plotted on the same axes (Figure 4-31):

- *Planned spend (PV)* – generated through development of a detailed logic linked, resource loaded schedule, converting deliverable progress into cost spent.

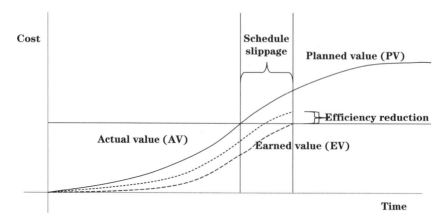

Figure 4-31 Defining earned value

➤ *Actual spend (AV)* – generated from the actual cost expenditure as the project progresses.
➤ *Earned value spend (EV)* – based on the original budget and schedule, each deliverable is assigned a value which is earned as the deliverable is progressed.

This graphical representation of progress can then be interpreted:

➤ *Cost axis* – If earned value (EV) is less than actual value (AV) then the deliverables are being generated at less than 100% efficiency (less deliverable progress per unit cost) and costs are likely to exceed budget.
➤ *Schedule axis* – If earned value (EV) is less than planned value (PV) then the deliverables are being generated slower than planned (less deliverable progress per unit of schedule) and the schedule is likely to be delayed.

For example, the design of a new manufacturing facility involves the development of 200 specifications for key equipment. These are to be developed by an external supplier on a target-reimbursable basis ($100,000 target cost). The majority of the equipment is on the project critical path and so an earned value method of progress measurement is to be used. The specifications are required to be completed to a specified quality standard within 6 months. The supplier is to receive monthly payments against approved timesheets. The earned value system will only allocate 'value' to a specification which is 100% complete. Table 4-15 shows the monthly progress plan agreed (PV), the interim payments made (AV) and the actual specifications completed (EV).

At this stage the Project Manager wanted to forecast the anticipated final cost for this contract and he did so by extrapolating the cumulative efficiency at the end of month 5. In addition he wanted to forecast the anticipated end date. He therefore followed the following process:

1. Convert all figures into financial value and make them cumulative rather than in month so that an 'S' curve can be plotted. For example, if 200 specifications has a financial value of $100,000, then 1 specification is equivalent to a financial value of $500.
2. Calculate the cumulative earned value in the same units (Table 4-16).
3. Plot the 'S' curve showing all three lines (Figure 4-32) and evaluate cost and schedule adherence.
   ➤ The EV line is below the AV line showing that the production of specifications is less than 100% efficient. The external supplier is spending more time developing each specification than they had planned.

Table 4-15 Example earned value progress measurement system (1)

Month	Planned production of specifications/month	Actual monthly supplier payments ($)	Actual delivery of completed specifications
1	15	0	10
2	25	20,000	20
3	40	20,000	40
4	65	20,000	42
5	30	20,000	28
6	25		

Table 4-16 Example earned value progress measurement system (2)

Month	Cumulative planned value ($)	Cumulative actual value ($)	Cumulative earned value ($)
1	7,500	0	5,000
2	20,000	20,000	15,000
3	40,000	40,000	35,000
4	72,500	60,000	56,000
5	87,500	80,000	70,000
6	100,000		

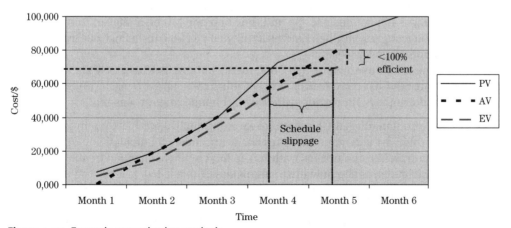

Figure 4-32 Example earned value analysis

- The schedule is also slipping as shown by the difference between the EV and PV lines. Assuming that the efficiency of production remains the same then the overall delay will be approximately a month.
4. Calculate forecast cost to completion:
   - Efficiency by month 5 = EV/AV = 70/80 = 87.5%.
   - There are 60 deliverables left to complete, so cost to do this at same efficiency = 60 × $500 × 100/87.5.
   - Therefore cost to complete deliverables = $34,000.
   - Total forecast cost = $80,000 + $34,000 = $114,000.
   - A cost overrun of $14,000 is forecast.
5. Calculate forecast schedule to completion:
   - Planned schedule required approximately $70,000 EV by month 4.
   - So at month 5 the schedule has slipped by 1 month.
   - Schedule slippage = PV/EV = 87.5/70 = 125%; its taken 25% more elapsed time to do current work.
   - Total schedule to completion = 6 months × 1.25 = 7.5 months.
   - A schedule slippage of 1.5 months is forecast unless mitigating actions are possible.

In this case the Project Manager was able to negotiate with the external supplier to increase resource availability so that there was no further schedule slippage, however the cost overrun could not be prevented and required the use of project contingency.

Earned value is a powerful progress measurement, forecasting and control tool. However, it is the set-up of the earned value system within the control plan that is the most crucial element. Not every aspect of a project scope will need earned value and a good Project Manager will select the most appropriate parts of a project where this methodology will deliver value-add to the control strategy.

Change control

A key part of maintaining control is ensuring that the required scope is delivered in terms of its quantity, quality and functionality. The usual way to do this is to define a change control process and to use formal change control request forms which may be customized to align with key project drivers. It is then important that all requested changes are logged and the response against each noted.

The biggest issue for a Project Manager when a change request is made, is actually confirming that it is indeed a change. Changes impact cost, scope and time and need to be analyzed for their impact on all three. However some requests may not be changes at all, just reasons for a cost overrun, or a schedule slippage, or a rectification of a scope error.

For example, a business improvement project had identified that five new SOPs would be required to support the improved process. The cost estimate of 40 hours to be spent over 2 weeks was approved. During the project an additional SOP was identified and a change request was made:

- The change requested an additional 10 hours and an additional week.
- The 10 hours could easily be accommodated from the contingency budget.
- The additional week impacted the critical path as the final SOP was needed to start formal training in preparation for the launch of the new business process.

The Project Manager authorized:

- An additional 8 hours taken from the contingency budget (based on the original budget – 40 hours divided by 5 SOP's = 8 hours per SOP).

The resource to work full-time on the SOPs so that the extra SOP could be completed within the original 2-week time-scale.

The extra 2 hours was actually an anticipated overrun rather than a change and as such should not be taken from contingency.

What is change?

Change is defined as any issue or decision which alters the design or delivery basis established by the control plan.

Who raises a change?

Change (which is a normal project occurrence) can be initiated by any member of the Project Team, but only progressed with approval from the Project Manager.

Forecasting

The culmination of any control plan is a test of whether the Project Manager is adequately able to forecast the project outcome with the selected methodologies, tools and detailed plans in place.

Without actually forecasting the tools are merely monitoring the status quo. A Project Managers job is to continually review the horizon: the project outcome, and to increase the certainty with which he can forecast it. A Project Manager should:

- Assess all elements of the control plan and estimate the end of project situation based on knowing what has been done. This means assessing the potential for:
  - Project success.
  - Adherence to cost and schedule plans.
  - Contractual issues.
  - Scope issues.
- Review trends for possible indicators of issues which could impact the project outcome:
  - Performance metrics.
  - Team performance.
  - Supplier performance.
  - Senior stakeholder interactions.
- Be able to assess whether the project is in control and whether it is likely to remain so.

Project review strategy

A project review strategy is an integral part of any control plan. It defines the methodology for project performance management:

- The metrics for assessment of performance.
- The target levels of performance and their link to project success.
- The way in which performance data will be collected and communicated.
- The way in which performance data will be used to control the project outcome.

The project review strategy should link in with the defined control strategy and the other elements of the project control plan.

Project performance review

Project performance is reviewed at three distinct levels in order to adequately cover all typical root causes of poor performance:

- *The project* – sometimes there is a fundamental flaw in the project – perhaps a misalignment between project constraints of cost, scope and time, a risk profile which cannot be mitigated or an inappropriately placed contract.
- *The team* – sometimes teams do not perform as required.
- *The individual* – sometimes individuals do not perform as required.

Project Performance Metrics

In developing any metric the one thing to remember is that 'measures determine behaviours'. A project scorecard (Appendix 9-11) would usually focus on the highest level of performance management, develop metrics to identify current performance and relate these to the overall project goals so that a forecast of project outcome can be made. For example:

- *CSF traffic lights* – the status of each CSF as a risk rating denoting whether the CSF will be successfully achieved.
- *Progress against schedule* – this is sometimes converted to a percentage progress for ease of comparison to plan and can incorporate an earned value metric.
- *Time* – to complete a specific project stage versus typical benchmark times for similar projects.
- *Progress against cost plan* – this can also be viewed as a percentage spend against a planned spend and an earned value spend.
- *Number of deliverables* – tracking specific deliverables against plan in terms of interim and final status linked to a quality target.
- *Number of changes* – tracking the total number of design or contract changes.
- *Safety or quality specific metrics*- tracking specific areas of scope which may be of particular relevance within a project. For example, the number of accidents or the number of quality issues raised.
- *Close-out metrics* – tracking outstanding areas of scope and also any benchmark metrics (overall project schedule and cost adherence or for specific phases of the project).

A project scorecard has three distinct audiences and it has to simultaneously meet all three:

- **Project Manager:** Although the project scorecard is usually developed by the Project Manager he can also be considered its main audience. This is a self-assessment to demonstrate that the project is in control.
- **Project Team:** A project scorecard can give a large diverse team a common understanding of the current status of both progress and issues and allow team members to see the impact of their performance. For this latter reason metrics need to be chosen very carefully so that it drives appropriate team behaviours.
- **Project sponsor and customer:** A project scorecard can support the right level of communication to key project stakeholders. However, care needs to be taken with selecting metrics that the stakeholders will understand and react appropriately to.

For example on a recent plant shut-down project the overall project scorecard was very strongly linked to the construction activities as this was the major area of scope which required control (Figure 4-33). This focused the team on key areas and reinforced the importance of plant start-up on time, whilst also maintaining a good safety record.

Generating a scorecard which supports all levels of performance (project, team and individual) is the ideal situation although not always possible. Having internal and external scorecards is sometimes more appropriate. For example, a project to deliver an increase in sales performance has two levels of scorecard:

- An external scorecard which is issued to the sponsor, customer and other senior stakeholders. This highlights the potential for benefits realization and drives appropriate stakeholder behaviours in terms of delivering business change.
- An internal scorecard which is available to all Project Team members. This highlights key project scope metrics which drive appropriate team and individual behaviours.

Team performance metrics

Team performance metrics are usually linked to specific work packages which the team have responsibility for. Although where a critical path of success methodology has been used there should also be some general performance indicators. For example, a team within a major product development project have the responsibility for packaging a new consumer-based product. Their performance is measured with respect to the completion of a set of deliverables with both time and cost targets. Overall the project CSFs have highlighted the following:

- The need for a product which fills a very specific niche. Customers in this niche have a very low tolerance to non-value add, both ethically and economically.
- A product with a low cost to customer and therefore low production cost.
- The need to meet the summer market when the product is most likely to be needed by the target consumer.

The team notice board would highlight a 'smiley face' against each CSF, each week, in terms of whether the team had achieved something which supported meeting these CSFs. For instance during the early design phase the team focused on developing minimal packaging which was environmentally friendly – a combination of recycled and biodegradable. This 'scored' them two smiley faces.

Team measures determine team behaviours and a set of aligned team performance metrics can positively support the team pulling together and meeting their shared goal.

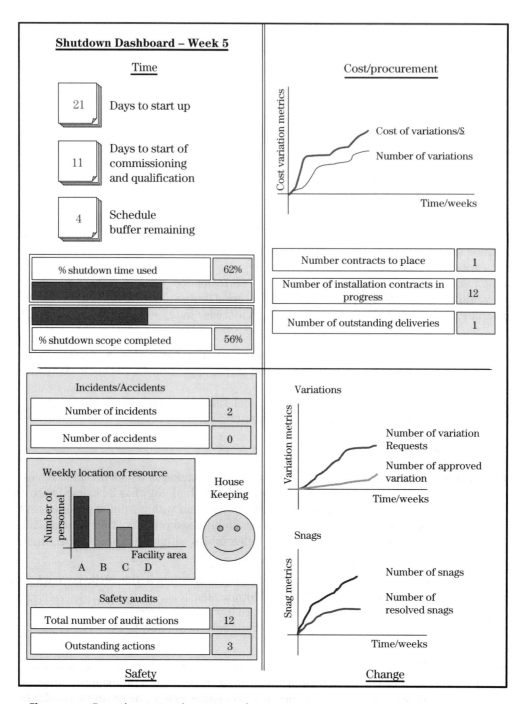

Figure 4-33 Example construction scorecard

Individual performance metrics

At an individual level performance metrics are usually viewed in the context of the person rather than the project. However it is important that the expectations a Project Manager has of a team member aligns with the expectations that the organization has. In other words, the way an individual is managed and developed on a project should meet his personal development plan goals. If they are in contradiction then the resulting confusion, in how a person is supposed to behave, will result in poor performance.

For example, if an individuals' development plan has highlighted a major gap in the way that he communicates with customers, then it may not be appropriate to give that person a major liaison role between senior stakeholder groups. In this case a developmental action may be to ensure that the team member is involved in customer interactions and sees appropriate modelling behaviours from those leading the activity.

An individuals' project scorecard should therefore be a sub-set of the overall development plan for that individual and both should align with each other. It should detail a balanced view of performance in terms of:

- Skills used and developed.
- Knowledge generated and shared.
- Behaviours displayed.
- Activities completed.

It is easy to focus solely on an individuals' activities – the tangible evidence of performance – but this does not always tell the whole story. For example, on an engineering project a lead mechanical engineer was completing all required activities on schedule, however it was clear that he was not integrated into the team. He was not liked because he behaved in a way which ensured that his work was completed at the expense of overall team goals. A development plan should highlight the importance of team working and meeting shared team goals.

Collecting performance data

Performance data can be collected in a variety of ways:

- Formal or informal meetings.
- Formal or informal written communications.
- Continuously, ad hoc or at specific milestones.

They can involve a variety of project stakeholders depending on the goal of the performance data. This mixture often leads to Project Managers generating a review strategy as a matrix or type of data versus stakeholders involved in the data collection (Figure 4-34).

Using performance data

Performance data should be a key input to both the internal and external communications plan. On the basis that measures do determine behaviours then they first have to be measured and then communicated.

Type of performance data	**Formal**	Project progress data against plans gathered frequently in line with the project roadmap (linked to project phases and schedule milestone). Gathered during formal teams meetings and from formal project reports or progress measurement systems.	Project audits at stage gates in the project roadmap or at critical milestones in the project schedule. Gathered by an independent project management resource able to quickly interpret high-level data and assess performance – based on metrics and observed behaviours.
	Informal	The normal behaviours of a high performing team would be a continuous assessment and exchange of performance metrics. A culture of informal discussions and a focus on meeting project, rather than solely individual, goals drives performance management into the team culture.	Informal knowledge exchange with senior stakeholders such as sponsors, end users and customers is a way to generate a clear view of expectations versus progress. In addition the Project Manager can use individuals within this audience as an informal project mentor, testing project scenarios and getting advice on direction.
		Internal to project	**External to project**
		Stakeholders involved in data collection	

Figure 4-34 Project review matrix

Typical communication methods are:

- Team meetings, reports, email bulletins and notice boards.
- Use of the company's intranet for wider communication of project progress to the whole organization or similarly use of an organization-wide newsletter.
- 1-1 meetings at all levels within and external to the project.

In addition a Project Manager has his own way of using project review data:

- To forecast whether the project will meet its intended goals.
- To provide guidance and direction to the Project Team.
- To ensure that all project activities are in alignment and if not, to support realignment.
- To highlight achievements and best practices and to recognize and reward good team performance.
- To identify problems and to manage poor team performance.

Control plan case study – construction project

To illustrate the key points from this chapter on control planning an extract from the PDP for a construction project is used.

Situation

A new headquarters for an organization based in mainland Europe was due to be built in a number of phases:

➤ **Phase 1**: Infrastructure for the campus including all temporary roads and facilities included in later phases.
➤ **Phase 2**: Offices for the corporate division and all marketing research departments.
➤ **Phase 3**: Offices for the various product divisions and conversion of all temporary facilities into final versions.

Phase 1 had proceeded to plan and the organization was so satisfied with the outcome that they decided to use the same main contractor. However phases 2 and 3 were more complex in terms of the build structure and architecture. The time-scale was also a more significant challenge as people needed to be moved into the offices more quickly than originally expected due to a recent company expansion.

For phase 2 (to be quickly followed by phase 3) the development of a control plan was particularly critical in managing a vast array of external suppliers, significant cost and schedule constraints and a large, complex construction site (aiming to achieve benchmark safety and quality metrics).

Control plan

The Project Manager, with support from other project management expertise from within the company, completed the development of a thorough control plan. Due to the type of project the majority of the Project Team consisted of external resources with a few internal facility management experts and a small internally resourced project office (accountants, planner and administration).

Risk strategy

The Project Manager used a critical path of risks methodology to highlight the high-priority risks and current project issues. The CSFs were taken from the WBS and followed the project CPA:

➤ *CSF 1*: Appropriate facility quality and operability (understand customer needs).
➤ *CSF 2*: Appropriate internal project management (manage risks and achievement of the project goals).
➤ *CSF 3*: Design completed (appropriate cost, time, quality and quantity).
➤ *CSF 4*: Construction completed (appropriate cost, time, safety and quality).
➤ *CSF 5*: External suppliers perform (project management, control and quality).

These supported the achievement of the project goal: *phase 2 handed over on-time and within budget to a satisfied customer.*

Risk management

A risk was defined as a potential issue that could significantly impact the achievement of the project objectives if it occurred. Once a risk was identified it was assessed:

- ➤ Likelihood that this risk will actually occur – low, medium or high.
- ➤ Potential impact if the risk did occur – low medium or high.

Then two types of plan were developed:

- ➤ *Mitigation plan* – an action plan to eliminate the possibility of the risk occurring.
- ➤ *Contingency plan* – a back-up plan for use when the risk has occurred, which minimizes impact on the project.

During the planning phase 37 separate risks were identified of which approximately half were rated as amber (Figure 4-35). Due to resource constraints only the amber risks with a 'high' score in either probability or impact were mitigated (11 mitigation action plans were developed and resourced).

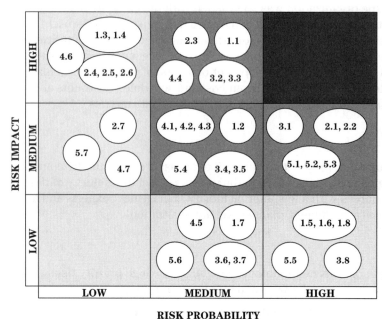

Probability		
	Score	**Comment**
Low	1	Very unlikely to occur
Medium	2	An 'evens', chance that this could occur
High	3	A high possibility of occurrence
Impact		
Low	1	Minimal impact on project goal
Medium	2	Significant impact project goal difficult to achieve
High	3	Critical impact, unable to achieve project goal

Red risks

Amber risks

Green risks

Figure 4-35 Case study risk matrix

A critical path of risks was developed in order to focus on those CSFs with the greatest risk profile (Figure 4-36). This was to be reviewed each month to check that mitigation plans were successful and that the risk profile trends did not move towards 'red'.

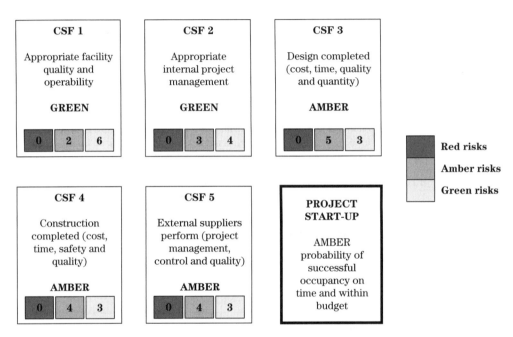

Figure 4-36 Case study path of critical risks

Critical issue management

A critical issue is a current situation which has the potential for significantly affecting the achievement of the project objectives. During the planning stage two critical issues were identified (Table 4-17). Each issue has a recommended action plan. Additional critical issues may be identified as the project proceeds (because risks can turn into issues). A critical issue log will be kept as a part of the control plan.

Table 4-17 Case study critical issue log

Critical issue log				
Project:	Construction project	**Sponsor:**		Pierre Bleu
Date:	Month 1 year 1	**Project Manager:**		Michel Noir

Issue number	Issue	Responsible	Status
A.1	Office specification does not meet senior end user requirements	Facility management representative (FM)	In progress

Statement of issue
A recent survey conducted by the corporate facility management team has identified that the office specification to be used in the new facility (one which is noted as standard within all European offices of similar major companies) is under challenge by the majority of divisional leaders. This is due to the increased level of open plan accommodation, the proposed occupancy levels and the size and furnishing of offices. There is also concern regarding the new policy to have only networked printers in open plan areas.
Objective
To gain senior stakeholder buy-in to the new office standards and policies.

Action plan	Action by	Target date
1. Ensure that the campus and accommodation layout drawings were available for review	FM	Month 3
2. Construct a sample office for review by all senior stakeholders	FM	Month 6
3. Consider the confidentiality needs expressed by some divisions with regard to open plan working	FM	Month 6

Apart from this live issue all other CSF 1 risks are GREEN.

A.2	Excavation work plan impacted by inability to get plant hire confirmed	Construction site representative (CM)	In progress

Statement of issue
The civil works tender is in progress to enable excavation to commence as soon as planning is completed, however the large machinery required cannot be sourced and this will cause a major delay to the whole project.
Objective
To get the machinery on site ASAP and commence work as per plan.

Action plan	Action by	Target date
1. Contact architect for contacts	CM	ASAP
2. Contact construction consultant for contacts	CM	ASAP
3. Maintain pressure on usual supplier	CM	ASAP

Apart from this live issue all other CSF 4 risks are either AMBER or GREEN.

Contract strategy

The scope was divided into 36 procurement packages (Table 4-18) to give the organization the best possible cost outcome:

- One construction management contractor providing overall management of the design and build scope.
- One design consultant providing the architectural design and design basis for the main sub-contracts.
- One civil works contract, 10 building shell contracts and 7 internal building contracts.
- Nine building infrastructure contracts and 7 external campus infrastructure contracts.

Table 4-18 Case study contract plan

Planning Toolkit – Contract Plan							
+Project:	Construction project			**Sponsor:**	Pierre Bleu		
Date:	Month 1 year 1			**Project Manager:**	Michel Noir		
Procurement package	**Contract style**	**Milestone dates**					
		Resource listing	**Tender package**	**Resource selected**	**Contract placed**	**Contract completed**	
Construction management contract	Target reimbursable fees versus man-hour estimate	Single source Use phase 1 supplier but check target price	Month 2	Month 2	Month 3	Month 24	
			⇒ Scope of services ⇒ Contract plan	n/a		Cost and schedule adherence	
Design consultant	Target reimbursable fees versus a schedule of deliverables	Single source Use phase 1 supplier	Month 2	Month 2	Month 3	Month 10	
			⇒ Scope of services ⇒ Concept design	n/a		Quality and operability of final design Cost and schedule targets Delivery of the technical specification	
Civil works contract	Bill of quantities	Single source Use phase 1 supplier	Month 3	Month 3	Month 4	Month 10	
			⇒ Scope of supply ⇒ Technical specification	n/a		Cost and schedule adherence	

(continued)

Table 4-18 (Continued)

Planning Toolkit – Contract Plan							
Project: Construction project				**Sponsor:** Pierre Bleu			
Date: Month 1 year 1				**Project Manager:** Michel Noir			
Resource required	**Contract style**	**Milestone Dates**					
		Resource listing	**Tender package**	**Resource selected**	**Contract placed**	**Contract completed**	
Building shell sub-contracts (10)	Fixed price and agreed variation SOR	All phase 1 suppliers (those selected and those pre-qualified but not selected)	Month 5	Month 6	Month 7	Month 19	
			➤ Scope of supply ➤ Technical specification ➤ Document requirements	➤ Reliability ➤ Safety ➤ cost		Cost and schedule adherence	
Internal building sub-contracts (7)	Fixed price and agreed variation SOR		Month 8	Month 9	Month 10	Month 22	
			As above	Quality and cost		Cost and quality adherence	
Building infrastructure sub-contacts (9)	Fixed price	Use design consultant to develop list of suitable suppliers	Month 10	Month 11	Month 12	Month 24	
			As above	Quality and cost		Cost and quality adherence	
External campus infrastructure sub-contracts (7)	Fixed price		Month 10	Month 11	Month 12	Month 24	
			As above	Quality and cost		Cost and quality adherence	

The highest priority for the project was cost (capital and lifecycle). However, the customers of the campus were either senior leaders in the organization or key divisional leaders developing future products. This meant that it was also important that the quality of the specification was appropriate. These drivers were incorporated into the contract plan (both in terms of selecting the suppliers and in choosing a contract style and contract performance metrics). The contract plan (Table 4-18) was used as a milestone tracking plan to support progress measurement and overall control of the project. However, as not all the dates were critical this was supplemented with the schedule risk assessment and review of the CPA.

Control strategy

Introduction

Appropriate project control tools were used to monitor progress against project objectives. These were cascaded into all major contracts so that there was an aligned project management methodology to control the outcome.

Schedule management

There was a cascade of logic linked, resource loaded Gantt charts based on the master schedule. Each supplier generated and owned their owned schedule. Based on a schedule risk assessment the civil works contract was deemed high schedule risk and a 4-week buffer was placed in the schedule to protect the other works (fixed price contracts). A further schedule buffer was placed just before occupation to protect the final handover date. This was only a 3-week buffer and is largely based on the risks of getting the lifts (elevators) to site and installed on time. The two buffers were controlled by the Project Manager.

Cost management

A standard cost plan was used to control costs (Figure 4-37) and the development budget has been separated from the contingency budget. As external suppliers are being managed by the construction contractor that supplier will own a part of the cost plan, although they will have no visibility of the internal organization costs or the contingency budget.

The total approved cost for the project is $30,000,000 with 4.5% contingency and 5% development budget. This was to be used as planned by the contingency run-down chart (Table 4-19).

Table 4-19 Case study contingency run-down chart

Item	Contingency (%)	Development (%)	When run-down is scheduled
Construction management contract	0.5	None	During the contract
Design consultant	0.5	None	During the contract
Civil works contract	0.5	2	Anticipated excavation problems
Building shell sub-contracts (10)	1	1.5	At end of contract
Internal building sub-contracts (7)	1	1	At end of contract
Building infrastructure sub-contacts (9)	0.5	None	At end of contract
External campus infrastructure sub-contracts (7)	0.5	0.5	At end of contract
Total	**4.5**	**5**	

The cost plan and contingency run-down were reviewed on a monthly basis. All internal organization costs were considered 'below the line' and not subject to capitalization. As such, they did not appear in the cost plan.

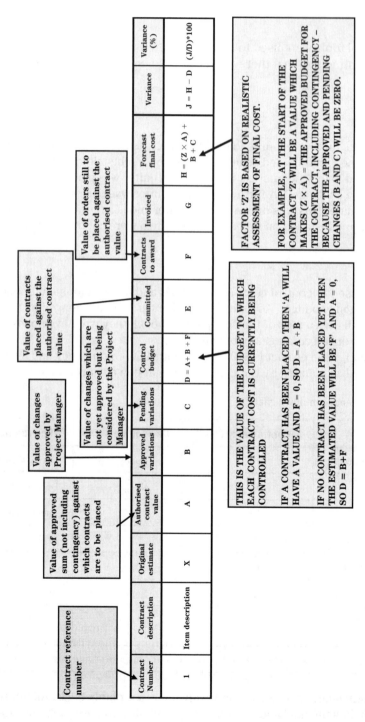

Figure 4-37 Case study standard cost plan template

Progress measurement and forecasting

There were nine main metrics to assess progress and forecast the project outcome:

1. Tracking Gantt chart with associated critical milestone plan – tracking the progress of critical milestone achievement.
2. Schedule buffer run-down – tracking the use of the 7 weeks of schedule contingency.
3. Cumulative cost spend versus plan – tracking the value of contract placement, cost accrual and cost expenditure versus plan.
4. Cost contingency and development run-down – tracking the use of the 9.5% cost contingency.
5. Design documentation earned value – tracking development of the design documents associated with each of the seven floors in the new building. Construction work could not commence without approved design documents.
6. Construction earned value – tracking the wall and floor build progress over the seven floors.
7. Construction resource histogram – tracking the level of resource on site as an indicator of level of activity and also linked to safety management.
8. Construction safety statistics – tracking total number of safe working hours as an indicator of appropriate site
co-ordination of a large number of contracts.
9. Construction housekeeping – tracking housekeeping and general build quality.

Scope management

The design consultant developed a design document matrix so that it was clear:

➤ Who owns each design document (in terms of the master copy).
➤ Who needs the design document to progress work.
➤ When design documents were ready for issue.

The construction contractor was responsible for the management of all project documents and co-ordinated with the Project Manager in terms of the development of tender packages. All changes were entered onto one master register and each was categorized as follows:

➤ Valid change of scope.
➤ Scope development.
➤ Contractual variation under dispute.

Any changes from the scope had to be approved by the Project Manager, who occasionally required sponsor approval before the change could proceed. At a project and work package level, all teams had a change control procedure in place for all critical deliverables once approved. This control plan, and all others below it, was also subject to change control once approved.

Reporting

Project reports were used to report progress against objectives. The project reporting strategy was cascaded so that all major external contracts reported progress in line with project needs.

The system of reporting proposed (Figure 4-38) was based on a one-page high-level project report. The contents of the report were as follows, with the level of detail as applicable to the level of the report:

➤ Key activities or achievements completed in month and those planned for next month.
➤ Milestone or objective tracking.
➤ Forecast of project or work package success.

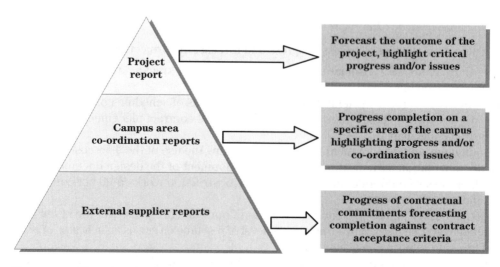

Figure 4-38 Hierarchy of reports

Review strategy

Due to the complexity of the project organization and the large amount of external suppliers, the review strategy considered a matrix of reviews (Figures 4-39) in order to manage performance.

		Project Team	External suppliers	Independent reviewers
Review type	**Formal meeting**	Weekly co-ordination meeting Monthly progress review (including risk, cost, schedule, change review)	Weekly co-ordination meeting for all external suppliers Weekly progress meeting for each supplier on site	Formal quarterly IPR
	Formal 1-1	All team leaders have formal monthly 1-1 meetings with the Project Manager	Each external supplier Project Manager will have a formal monthly 1-1 with the Project Manager	Both the sponsor and customer will have a separate 1-1 with the Project Manager
	Informal	All team members are encouraged to maintain informal contact with each other Project Manager to maintain regular contact	Project Manager 'walk-arounds'	External technical or project management advice upon request
		Project Team	**External suppliers**	**Independent reviewers**
		Review participants		

Figure 4-39 Case study review strategy

The Project Manager wanted to encourage regular informal contact between team members in order to allow information to flow constantly around the project – supporting alignment and a team culture. In addition he used key 1-1's with the project management professionals in the external supplier organizations to maintain an overview of certainty of success in all contracts.

All elements of the project were available for an independent project review (IPR) of performance once the planning phase was completed. A key milestone where an IPR can add benefit to the business is immediately prior to the start of construction, when all project strategies are in place. This IPR was formally requested to support the project performance management strategy.

Specific external reviews can address project management or technical issues. The Project Manager recognized and integrated these into the strategy recognizing that such review were an opportunity for the Project Team to share learning, both within the project and also across the organization.

Conclusions

Although phase 2 of the project went through an extremely 'uncontrolled' stage, this was in-part due to the lack of use of the control plan by the Project Manager, requiring project management intervention. The regular independent review conducted by a senior Project Manager from within the organization identified at a very early stage in the construction that progress was behind schedule, costs were escalating and external suppliers were not being adequately co-ordinated or controlled. The root cause was poor project management at the overall level and at the supplier level.

As a result a new Project Manager was appointed and all parts of the control plan were implemented, bringing the project back into control. Phase 2 finished on time with an 8% cost overspend, no major safety incidences and a satisfied customer (the employees who were using the campus).

As a result of the issues during phase 2, phase 3 was kicked off with a stronger project management team and a different main contractor. The change of contractor was necessary due to the project management issues which were never entirely resolved during phase 2. They were only controlled through heavy intervention and detailed task management. In effect the organization was paying for project management *and* having to do it themselves!

Phase 3 was planned and delivered successfully and the organization learnt a lot more about project management, which they were able to use in future business projects.

Key points

The aim of using this particular case study was to demonstrate that:

- It's no good having a control plan unless you use it.
- There is always value in having an independent project review to highlight the issues which a Project Manager can get 'blind to' when in the middle of a complex fast-track project.
- Planning needs to cascade into the supplier contracts so that delivery plans for all contracts align with the overall project requirements.

Handy hints

No control plan means no 'control'

Obvious though it sounds, the chances of success are increased by firstly having a control plan. Then you need to actually use it during delivery!

Risks change, risks move and risks sometimes don't happen

Too often robust risk assessments are completed at the start of project planning but never reviewed again prior to delivery. For a major project (when the planning can take some weeks) it is important that risks are reviewed, mitigation plans are revised as necessary and that the right risk tools are in place for the start of delivery.

External suppliers need to be aligned and managed

For some reason organizations seem to believe that a supplier has only their contract on their radar, with the consequence that managing them is deemed a minor project management activity. Suppliers are businesses with many customers and whether or not they have a good level of project management capability in-house they still have different goals. Sometimes this can be managed via alliances or other partnership relationships, but whatever the contractual link, suppliers need to be aligned and managed.

Understand the control needs of your project

How a small business improvement project is controlled will be quite different from how you control a major multimillion dollar capital engineering project. Methodologies, tools and processes need to be selected with a specific project in mind so that they meet specific project control goals.

Push for independent reviews

Having a peer review of your project is a positive thing! Any Project Manager knows that a key part of performance management is getting a fresh view on a situation and sometimes stating the obvious. A review strategy is key to identifying how you will get this support and should be a planned, positive process.

Keep reviewing certainty of success

During control planning it is important that you keep assessing how the control plan will support success by increasing the certainty of a positive outcome. In effect, the control plan is the tool you will use during project delivery to keep making this assessment; so it needs to be composed of the right tools, be aligned to the project needs and above all be useable and easy to communicate.

And finally . . .

A robust project control plan:

- Articulates the project control needs in terms of how the project will be delivered in control – it considers the control of all processes necessary to deliver the project.

- Has a link to the project business plan – ensuring that the business benefits will be enabled.

- Has a link to the project set-up plan and its requirement to use organizational resources effectively.

- Is built on robust processes and effective project control management.

- Enables forecasting of the project outcome and supports the increase in the certainty of outcome.

5 The project delivery plan

Once the three tiers of planning activity have been completed, encompassing the 11 planning themes, a series of deliverables will have been developed. Collating these into a 'formal document' is then required. This chapter considers the form of the document and also the techniques for generating it.

Why we need a formal document

The focus of all the planning activities described in this book has been on the development of an integrated and robust Project Delivery Plan (PDP). This is a crucial document which sets out exactly how the project will be delivered so that it achieves both project and business objectives (Figure 5-1).

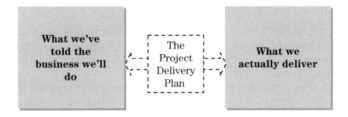

Figure 5-1 Why we need a formal PDP document

A lot of experienced Project Managers may tell you that they 'don't plan' when what they really mean is that they don't write it down. They plan themselves but exclude all the benefits of engaging with the vast array of project stakeholders. Without a formal PDP there is no plan, and then the project delivery strategy is just 'hoping for good luck'. Project management is to a large extent about the management of people and the plan is an effective way to work out (and document) the best way to have the right two-way relationship with a broad variety of stakeholders: from customers to team members, sponsor to user groups.

The formal PDP should therefore be considered:

- A 'contract' between the Project Manager and the sponsor.
- A communication tool between the Project Manager and the Project Team.
- The basis for communications to all project stakeholders.

A PDP is specifically prepared for a project based on standard project delivery planning practices. Its aim is to ensure that strategies specific to the needs of that project are developed, implemented, communicated and constructed around best practice project management.

- It defines the project purpose and objectives, the Project Team organization (with individual responsibilities) and the delivery strategies required to ensure that the outcome is successful and meets the business needs.

➡ It is a key communication tool for the organization – all members of the Project Team, the sponsor and all other stakeholders within the business.

The PDP is a 'live' document which should be updated at regular intervals. This is required to ensure that as strategies are developed and refined, they are aligned in their achievement of the project objectives and business benefits. The initial PDP should be developed and approved prior to the start of the project delivery stage and then used by the Project Manager and Project Team throughout delivery. It should be considered as:

➡ A 'project procedures manual' which guides the Project Team, supporting decision making and information flows.
➡ A 'quality plan' aiming to deliver a 'quality project'.

A formal document, whether hard copy or electronic, is therefore necessary so that it can be used as described above. A part of this formality is the management of the review, approval and then subsequent changes to the document and any deliverables associated with it (Appendix 9-3).

Who develops the PDP?

Although it is the Project Manager's responsibility to develop the PDP, both the sponsor and the customer have specific goals from the planning process. These must be borne in mind by the Project Manager as planning progresses:

➡ *Sponsor* – ensure that the project is being planned to meet business needs.
➡ *Customer* – understand how the project will be delivered and then successfully integrated into 'business as usual' (BAU).

Both the sponsor and customer must therefore be clear about their role in the project delivery. In addition, the Project Manager will involve the Project Team in planning activities as appropriate and as discussed in the PDP lifecycle section: steady state.

The PDP lifecycle

During the development of the formal PDP, all elements of the planning lifecycle need to be considered for each of the 11 planning themes:

➡ **Start-up:** Determining the most appropriate tool and/or methodology for each planning theme.
➡ **Steady state:** Determining the appropriate technique with which to use the planning tool/methodology and then using it to develop the plan.
➡ **Close-down:** Continuity checking between the 11 planning themes.

Start-up

As shown in *Project Management Toolkit* (Melton, 2007) and Chapters 2–4, there are a variety of tools which can be used to develop a robust plan. It is critical that a Project Manager chooses the most appropriate tool for the specific project. Typical selection criteria include:

➡ Project size.
➡ Project type.
➡ Project complexity.

- Specific project drivers such as time, cost and regulatory.
- How the plan (the output from the planning tool) is to be used during the delivery of the project.

A good Project Manager will review his planning toolkit (Chapters 2–4) and select the simplest tool for the job. Alternatively some more complex tools can be adapted to support simpler situations.

A good example of the latter is the technique of using earned value to set a planning baseline. Earned value (page 200) can be used in large, complex projects to monitor the overall impact and potential variance of the progress of many activities. However it can also be used to plan and track the development of a small bounded set of deliverables. The project management skill is in understanding the methodology and then being able to apply it appropriately.

For the Project Manager the eventual use of any tool is a key criterion. For example, a strategic assignment may well have developed the schedule baseline as a logic linked Gantt Chart (page 191). However the customer is only concerned with two critical points in time and only activities which can impact these critical points. The Project Manager in this situation would select a schedule buffer technique to manage, control and report progress (page 195).

Steady state

Actually doing the planning is the role of the Project Manager and his team. The Project Manager should select the appropriate way to do the planning. Total team participation in every aspect of planning or planning in complete isolation are both inappropriate choices. The selection process must consider how each planning theme will be used during the delivery phase. For example, there is no point in asking the Project Team to develop the external communications plan if they are not going to deliver it.

At this stage a Project Manager would start to consider the level of empowerment which is appropriate for the delivery of the project and therefore ensure that the way the project is planned aligns with this (Figure 5-2).

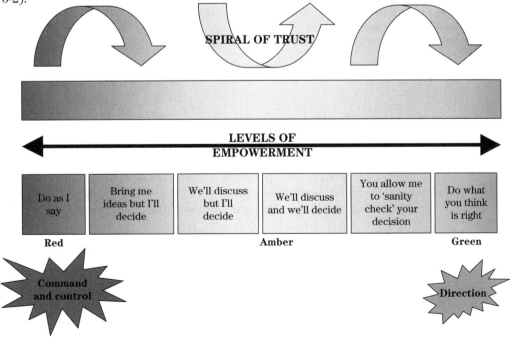

Figure 5-2 The continuum of empowerment

The level of empowerment needs to be appropriate for the specific situation (Table 5-1) and a Project Manager (denoted as 'PM' in Table 5-1) would typically use a variety of levels during a project.

Table 5-1 Typical planning techniques

Typical examples	Level of empowerment					
	PM delegates instructions	**PM invites ideas and makes decisions**	**PM discusses with team and makes decision**	**Joint decision using agreed project processes**	**Team decision based on PM guidance and review**	**Team decision within PM direction**
Planning forum	PM 'contracting' sessions with sponsor and customer	Project Team meeting (sub-project leads) with PM in chair	Project Team meeting with heavy PM facilitation	Project Team risk session with PM chair following structured process	Sub-project team meeting with sub-project lead in the chair	Sub-project team 'start-up' meeting
Planning technique	Checklist to structure data collection	Ideas brainstorm and selection criteria	Flowcharting and benefits mapping	FMEA, root cause and decision matrix	Checklist to structure options	Schedule brown paper technique
PDP deliverable	External communications plan	Project contract strategy	Benefit realization plan	Contingency plans	Project reports	Sub-project schedules
Delivery style	PM delegates activities to deliver key communications messages in line with plan	PM and Project Team deliver the contract strategy as planned	Project benefits reviews with team requiring PM decisions	Project risk reviews with team leading to joint agreed decisions based on structured process	Team owns the deliverable and is accountable for the output based on PM guidance	Team owns the deliverable and is accountable for the output

When a Project Manager behaves in a way which contradicts the level of empowerment then trust within the working relationships is broken.

For example, a Project Manager asks the design team to develop the design schedule themselves and summarize any critical milestones. However, upon receipt of the latter he asks for the detailed schedule and proceeds to amend durations and dependencies himself. The design team thought they were operating 'in the green' when actually the Project Manager had always intended to make the final decision (nearer to the amber/red part of the continuum in Figure 5-2). The team now distrust further instructions to 'do it yourself'.

However when behaviours align then trust is built up.

For example, a lead engineer makes it clear to his team that he has to formally 'sign off' a specific set of engineering calculations and that this must be worked into their delivery plan. During delivery of the plan the engineer ensures that the reviews occur, the team see this and trust is built.

A capable Project Manager will intuitively make these choices and select techniques which increase the probability that a robust plan will be developed and then delivered successfully.

Close-down

As described in chapter one (Figure 1-9), project delivery planning is an iterative process and a Project Manager will need to complete a series of planning continuity checks to ensure that all 11 parts of the plan align and support each other. Typical continuity checks include:

- Cost, schedule, resource and contract strategy inter-relationships. For example, does the cash flow profile match with the procurement schedule and the funding plan?
- Quantity, quality and functionality of scope link to benefits. For example, will the scope deliver the benefits?

Short case study

To demonstrate the use of the PDP lifecycle a short example is used.

A Project Manager has been asked to review the capacity of an aging production line in order to see if any further improvements can be made. The line is likely to be decommissioned in the next 5 years as the product is entering a mature phase and forecasts show a peak next year, with a decline thereafter. The management team therefore want to limit capital investment.

The Project Manager quickly reviewed the business case and his understanding of the business drivers helped him to conduct a quick and effective planning stage (Table 5-2) to deliver a PDP which was acceptable to the sponsor and project stakeholders.

Table 5-2 Case study – planning tool and technique selection

Planning theme	Planning tool	Planning technique	Continuity comment
Sponsorship	Contracting checklist Communications Plan (short-form)	1-1 with sponsor (Production Manager)	Check that sponsor contract agrees with vision of success
Benefits management	Benefits Tracking Chart	Review with sponsor vs sales targets	Review vs scope
Business change management	Sustainability Checklist Stakeholder matrix	Production line brainstorm	Cross-check with benefits plan Ensure that stakeholder plans encompass production management team
Project scope	Vision of success Scope matrix	Production line root cause analysis	Check that scope can be delivered with zero capital and within 4-week schedule
Project roadmap	Flowchart	Development by Project Manager	Check stage gates link with Communications Plan
Project organization	RACI	Development by Project Manager	Check alignment with resource histogram

(continued)

Table 5-2 (Continued)

Planning theme	Planning tool	Planning technique	Continuity comment
Funding strategy	Decision tree	Development by Project Manager	Check resource release approvals and revenue expenditure agreements
Contract strategy	None	None	No external suppliers necessary. Revenue items procured as a part of production BAU
Risk strategy	FMEA	Production line brainstorm	Check that all mitigation plans are in scope and schedule
Project control strategy	2-month Gantt Chart and 4-week delivery milestone chart (weekly) Resource histogram	Development by Project Manager and production line brown paper exercise for 4-week phase	Ensure that control plan will indicate 'out of control' status
Project review strategy	Milestone chart	Development by Project Manager	Check it matches sponsor expectations

Within 2 weeks a short-form PDP was presented to the production management team, approved and delivery commenced.

The project was completed within 4 weeks and the production line capacity was sustainably increased by a further 15%, thus supporting the short-term sales peak. No capital investment was required although revenue was expended (training, updated documentation, new visual control boards). At the after action review the Project Team identified 'good planning' as one of the major root causes for success followed by 'good teamwork'.

The Project Manager had identified at an early stage that the project needed the experience and expertise of the operators on the production line. He involved them appropriately in the planning as a precursor to their major role in delivery and then integration of the new ways of working into their own BAU.

PDP style, format and content

There are some simple rules to follow when developing the actual document:

Keep it simple!

Your PDP needs to be written in clear and simple language, so that the reader immediately understands what you intend to do. This is not a competition for the biggest document – it is a challenge for you as a Project Manager to effectively communicate how the project will be delivered.

Engage your sponsor and your Project Team

These are the people who will help guide the development of the PDP to a successful conclusion. Make use of their knowledge and experience to produce a relevant document.

Choose the most appropriate format for the PDP

There is no one prescriptive format for a PDP. An organization may have an in-house style and this is appropriate if based around the 11 planning themes. Two different styles/formats are presented:

➤ **A short-form PDP template** – appropriate for simple or short-term projects. A typical short-form PDP is very similar to the 'How?' Checklist (Appendix 9-2). It uses the 11 planning themes and allows the Project Manager to succinctly outline the plan and to reference key documents against each.
➤ **A detailed PDP template** – appropriate for complex or larger projects (Appendix 9-3). The template is separated into the three tiers of the planning hierarchy (Figure 1-7). It allows the Project Manager to succinctly summarize each part of the plan against the 11 planning themes whilst putting detailed planning deliverables in an appendix.

Both types of PDP format contain guidance on the type and level of content.

Choose the most appropriate planning tools

There are many planning tools available to use. Do not forget about the simple set of tools and techniques which are available to support creating the PDP. *Project Management Toolkit* (Melton, 2007) introduced a selection of tools and techniques which can support all planning activities – use them!

Write it down

As previously outlined (page 225) – if it is not written down it is only a rumour.

Projects, portfolios and programmes

Within project management the terms project, portfolio and programme are used to describe specific boundaries:

➤ **Project:** A bounded piece of work which is non-routine for the organization. It is not a part of BAU.
➤ **Portfolio:** A set of projects using resources from a common resource pool. These resources could be assets, people or funding, for example. There is a dependency between project resources which need to be used optimally.
➤ **Programme:** A set of interdependent projects working together to achieve a defined organizational goal. There is dependency of project output/benefits.

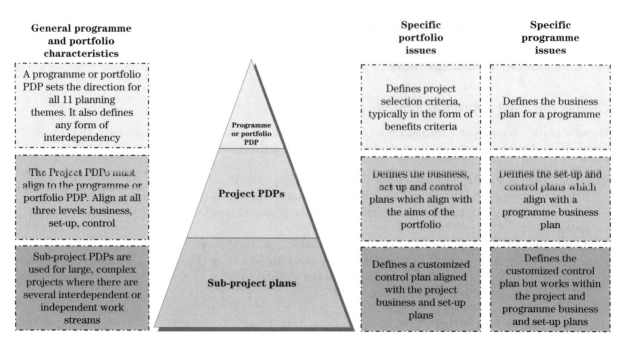

Figure 5-3 Hierarchy of PDPs

When either a portfolio or programme is defined it is appropriate that some high-level PDP is developed which sets the delivery principles for the overall portfolio or programme (Figure 5-3).

The PDP principles detailed in this book are appropriate to apply to all three, although there are some differences worth noting (Table 5-3).

Table 5-3 Projects, portfolios and programmes explained

	Project	Portfolio	Programme
Key characteristics	➤ SMART project objectives ➤ SMART benefit metrics	➤ A set of specific types of project within an organization	➤ A set of projects which together deliver a set of SMART benefit metrics
Managed by	➤ Project Manager	➤ Portfolio Manager and a team of Project Managers	➤ Programme Manager and a team of Project Managers
Project selection rationale	➤ Approved business case ➤ Organizational cost/benefit analysis	➤ Contributes to organizational goal ➤ Availability of required resources ➤ Prioritization of resources vs benefits criteria	➤ Contributes to programme goal ➤ Prioritization of project activities vs benefit metrics

(continued)

Table 5-3 (Continued)

	Project	**Portfolio**	**Programme**
Key planning goals	⬤ Delivery of a specified set of project objectives in control ⬤ Stakeholder engagement of the project activities and benefits outcomes	⬤ Delivery of a balanced portfolio delivering the highest priority organizational benefits ⬤ Optimal use of resources ⬤ Stakeholder engagement of the portfolio activities and benefits outcomes	⬤ Delivery of a specified set of programme objectives in control ⬤ Optimal flow of activities linked to flow of benefits ⬤ Stakeholder engagement of the programme activities and benefits outcomes

Example PDPs

The next three chapters contain full project or programme case studies. Each has been chosen to demonstrate a different issue and Table 5-4 summarizes these issues.

Table 5-4 Case studies explained

Chapter	Case study	Type	Description	Specific PDP issues highlighted
6	Capital engineering project	Project	The construction of a new (Greenfield) pharmaceutical manufacturing facility in Asia ⬤ Stakeholder engagement of the portfolio activities and benefits outcomes	⬤ Management of new product development and launch as it links to the facility ⬤ The business change issues regarding maintaining regulatory compliance after the facility had been handed over
7	Strategic consultancy assignment	Project	A project to develop a new business process within a services business	⬤ Using best practice planning tools on 'fuzzy' consultancy assignments
8	Organizational change programme	Programme	A programme of inter-related projects linked to the relocation of a hospital unit to a new building	⬤ The identification of additional projects and sub-projects within the programme upon which overall success was dependent ⬤ The need for an overall benefits plan

And remember . . .

- If it's not written down it's only a rumour!

- Keep it simple and make the plan appropriate for the project.

- A PDP is both a communication tool and a contract for delivery – it is not a technical project definition document. Write it with this in mind.

- A plan is only useful if you use it.

Case Study One: capital engineering project

This case study was chosen due to the complexity of the initial set-up. It was a project made up of different customer and supplier organizations, based on different continents with different views on how a project should be delivered. These various criteria impacted all aspects of project planning, the results of which are presented in a short-form project delivery plan (PDP) (Table 6-1) with accompanying documents. Only extracts from the PDP are presented to give an indication of the way the planning progressed.

Situation

A pharmaceutical manufacturing company had recently signed a licensing deal with an Asian manufacturing organization. As a part of this deal both companies had to work together to complete the development, approval and launch of a new drug which was to be produced in Asia. Due to the nature of the drug manufacturing process, a dedicated facility which met all local and international regulatory requirements was required. In particular the company wished to export the drug into North America and therefore full Food and Drug Administration (FDA) approval was needed.

An engineering project was sanctioned to build a new $40 million bulk manufacturing facility and a suitable Project Team was formed from a variety of organizations. This situation was novel for both main 'partners' and it was recognized that this would be a complex project to deliver. An initial set of challenges were highlighted and the assigned Project Director and his team of four Project Managers reviewed these challenges as a part of the high level planning process, with the aim of mitigating their impact on the project as much as possible.

Organizational interfaces

The two main organizations had an equal share in terms of decision-making power although the knowledge was divided as follows:

- Product development – Company X, based in Asia.
- General pharmaceutical manufacturing and regulatory knowledge and experience – Company Y, based in the UK.

Apart from these two organizations each company had a preferred main engineering contractor (the company who completed engineering design, procurement, construction, installation and testing for them). Due to the balance of knowledge, as described above, a decision was made to use both on the basis of specific knowledge and experience needs:

- Company X preferred to use a local engineering company which was a part of their 'parent organization' (Company A).

>> Company Y preferred to use a global engineering company which had successfully delivered similar facilities for them in the UK (Company B).

Before the project planning could even begin senior executives in all 4 organizations had to agree to the overall interface management and selection of key personnel:

>> Financial – Company X and Y.
>> Operational – Company X and Y.
>> Project management – Company B.
>> Compliance to pharmaceutical requirements – Company Y and B.
>> Conceptual design – Company B.
>> Engineering design and procurement – Company A and B.
>> Safety – Company X and A.

The agreed relationships between all four main companies are shown in Figure 6-1.

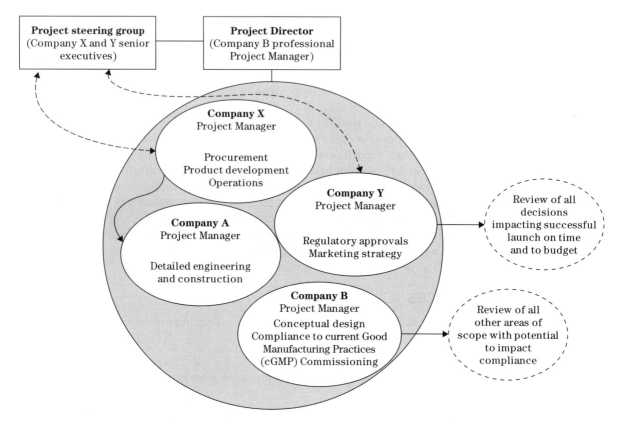

Figure 6-1 Relationship diagram

This high-level agreement gave the Project Director a baseline from which to develop an appropriate overall project delivery strategy and therefore the PDP.

Pharmaceutical practices

In order to develop, manufacture and launch a new pharmaceutical product, a company must meet specific requirements which demonstrate that the drug is safe and effective for patients. Company Y had extensive experience of meeting these regulations and already manufactured and sold drugs globally. Company X, on the other hand, had not previously manufactured a pharmaceutical drug of this type nor had they exported them previously. Therefore two issues were highlighted as potential challenges for the project:

Regulatory

In order to actually sell product, the facility needed to have regulatory approval. In this case, that meant approval from the FDA. Therefore there is a formal roadmap to be followed from New Drug Application (NDA) to formal approval which builds in various checks to ensure that there is:

- A robust scientific basis for the product manufacturing process: defining the aspects of the process or product that are critical to quality.
- A qualified facility: equipment, facilities and systems are proved capable of consistently performing as designed within their specified limits.
- A stable product produced consistently within specified limits (process and product).
- No risk to patient.

Compliance

It is imperative that compliance to current Good Manufacturing Practices (cGMP) is considered throughout the project and that the design builds in compliance to cGMP where possible. This can be

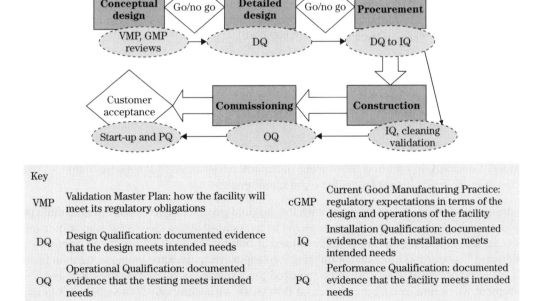

Key			
VMP	Validation Master Plan: how the facility will meet its regulatory obligations	cGMP	Current Good Manufacturing Practice: regulatory expectations in terms of the design and operations of the facility
DQ	Design Qualification: documented evidence that the design meets intended needs	IQ	Installation Qualification: documented evidence that the installation meets intended needs
OQ	Operational Qualification: documented evidence that the testing meets intended needs	PQ	Performance Qualification: documented evidence that the facility meets intended needs

Figure 6-2 Pharmaceutical project stage gate process

managed through the use of a specific project roadmap for pharmaceutical projects (Figure 6-2) which have clear go/no go points linked to cGMP checks.

Company Y and B felt that they had the experience to mitigate any issues in this area with the biggest challenge being to integrate cGMP and the regulatory milestones into all areas of the project.

Culture

It was clear from early discussions that culture required consideration:

➤ Company X and A worked within strong hierarchies with clear, segmented authority and little cross-functional working. Decision making was the responsibility of senior executives and teams were encouraged to excel within role or team boundaries.

➤ Company Y and B worked within team-based structures which were cross-functional and managed by a senior executive level. Decision making was delegated to the teams based on risk and teams were encouraged to maximize success for the company and to test boundaries if this delivered benefits.

In order to support Company X and Y the main structure needed to have clear levels of seniority and therefore all Project Managers were required to report directly to the Project Director. Each Project Manager was aware who had decision-making authority for specific decision types although they could organize their teams as they felt appropriate to the scope:

➤ Company A set up a functional engineering structure.

➤ Company B set up a task team structure and requested representatives from Company X's product development team.

Location

The management of Project Team location was a significant challenge in terms of both planning and delivery. There were to be five separate main locations for various elements of scope. In most cases work was in progress at many of the locations in parallel:

➤ Asia site 1 – Company X and A head offices.
➤ Asia site 2 – Company X manufacturing site.
➤ Asia site 3 – Company X research facility.
➤ UK site 1 – Company Y head office.
➤ UK site 2 – Company B head office.

Communications between the locations and associated time zones had to be managed so that decision making was not delayed. In addition there were times when the focus of the entire team had to be in one location, requiring significant travel for team members:

➤ Early design work required interaction with the original product development teams and review of pilot plant work, so the UK-based team needed to be located at Asia site 3 for a short period.

➤ Conceptual and front end design, which is aimed at building in quality, safety and operability, was completed at Company B head office but required input from all other teams at various times. Some of this could be done virtually and some required team members from Asia to be present.

➤ Procurement work required Company A and B to work with Company X to select suppliers and sub-contractors from all over the world, although the aim was to choose Asian companies where possible, given the cGMP requirements.

➤ Construction and commissioning were all based at Asia site 2 which was managed by Company X, although the construction site was managed by Company A. Company Y and B had a site presence throughout.

Business drivers

Due to the 'deal' which had been agreed between Company X and Y, the funding for all aspects of the project was extremely tight with little allowance for either escalation or contingency. The most critical aspect of this 'deal' was the delivery of the new drug to the market in time for peak use of the new drug. With regulatory timelines already fixed, this additional constraint meant that the project would need to be delivered approximately 3 months earlier than a comparable benchmark project.

Selection of overall Project Manager

Due to the size and complexity of the project the selected Project Manager was given the title 'Project Director' and she would have a team of Project Managers working directly for her from the various organizations. This Project Director was not from within either manufacturing organization although she had previously successfully delivered projects for Company Y. Senior management of the latter recognized that maintaining control would be the biggest challenge and so a professional Project Manager from a management contractor was selected.

Project management practices

The project management experiences within all four main companies were varied. Company X had little experience of major projects and Company A was very used to large but simple projects. Whereas Company Y and B were used to complexity and were strong in the use of current best practice project management. A major challenge would be to maintain equal control of all four parts of the project when the project management practices were so different.

Project Delivery Plan

Various extracts from the full PDP are contained in this section. The full PDP was a collation of an overall strategy and then four sub-project strategies which aligned with the general PDP principles. In this way, the Project Director could allow some variation in project management processes and tools but maintain overall control of the key elements which were fundamental to success.

Each separate PDP was finally checked and converted into a short-form document for ease of reference. The remainder of this chapter has extracted key parts of the PDP for the conceptual and front end design stage conducted by Company B.

Short-form PDP

Following completion and review of the PDP for the conceptual and front end design phase the Project Director compiled this short-form PDP (Table 6-1).

Table 6-1 The 'How?' Checklist: engineering project

PDP – Product ABC facility – conceptual and front end design					
Project Director:	Trish Roberts	**Revision:**	3	**Date:**	*Confidential*
Sponsor:	*Confidential*	**Status:**	Approved	**Copy no.:**	1
Executive summary					
This sub-project is a key part of the overall project. Links and dependencies to other sub-projects are described in a sub-project dependency chart (Figure 6-3). At this stage of the planning the dependencies between all projects appear aligned and in support of the overall project goal – product launch within 18 months.					
Sponsorship					
Who is the sponsor? (The person who is accountable for the delivery of the business benefits) A Vice President within Company X has overall accountability for the launch although he has a co-sponsor in Company Y. For this sub-project the sponsor issues have been delegated to the Project Manager within Company X on the basis that he will have overall responsibility for the operation of the new facility once it is built. Further reinforcing sponsorship is also being performed by the Project Manager from Company Y. Both of these Project Managers are senior executives who have the authority to make decisions impacting the facility operation and product approval. **Has the sponsor developed a communication plan?** There are to be no external communications from the Project Managers to stakeholders external to the Project Team. The Project Director will provide information to the steering group and they will communicate externally. Internal communications are the responsibility of the Project Director.					

(continued)

Table 6-1 (Continued)

PDP – Product ABC facility – conceptual and front end design

Project Director:	Trish Roberts	Revision:	3	Date:	*Confidential*
Sponsor:	*Confidential*	Status:	Approved	Copy no.:	1

Benefits management

Has a benefits realization plan been developed?
Benefits realization planning is within the overall PDP and this sub-project will contribute to it (Figure 6-4). A further check of the alignment between this project and the overall project benefits will be made at key decision points within the project roadmap (Figure 6-5).
How will benefits be tracked? (Have they been adequately defined?)
Benefit tracking has been defined within the overall PDP and will not need to be tracked separately within this sub-project.

Business change management

How will the business change issues be managed during the implementation of the project?
This sub-project has identified specific business change issues concerning the ongoing operation of the facility and developed a plan for their management during and after delivery of this project (Table 6-2).
Have all project stakeholders been identified?
This project has some very specific stakeholders linked to the technology transfer of the process from Company X research facility. There is also a strong link to the regulatory team. Therefore a Stakeholder Management Plan was developed to ensure that all appropriate engagement activities were progressed in support of the project goals.
What is the strategy for handover of this project to the business?
The output from this sub-project will be formally handed over upon completion. The outputs will be used within various sub-projects as outlined in the dependency map (Figure 6-3).

Project scope

Has the scope changed since Stage One completion?
This sub-project's main goal is to better define the scope so that detailed engineering and design can begin. A standard deliverables listing is available for conceptual and front end design and this will be used to develop interim and final deliverables. A cost estimate for the full project is also required.
Have the project objectives been defined and prioritized? (What is the project delivering?)
The main project objectives which should drive decision making (in terms of quality, quantity and functionality of scope) have been defined for the overall project and need to be used by all sub-projects to ensure alignment. A compliance hierarchy will also be generated within the conceptual design (Figure 6-8) for use in all subsequent design phases.

Project roadmap

What type of project is to be delivered? (For example engineering or business change)
This is an engineering project covering early design phases only (Figure 6-5).
What project stages/stage gates will be used? (Key milestones, for example funding approval, which might be go/no go points for the project)
A detailed roadmap has been developed so that compliance, regulatory, safety, operability and constructability checks can be completed at key stages (Figure 6-5).

(continued)

Table 6-1 (Continued)

PDP – Product ABC facility – conceptual and front end design

Project Director:	Trish Roberts	Revision:	3	Date:	*Confidential*
Sponsor:	*Confidential*	Status:	Approved	Copy no.:	1

Funding strategy

Has a funding strategy been defined?
Funding strategy is covered within the overall PDP and is being managed by the Project Director.

How will finance be managed?
The majority of the funding for the facility design and build is from Company X and as such their funding approval and finance management processes will be followed. As this is a design phase there is not expected to be significant expenditure unless some long lead items need to be procured early. The strategy for the early release of capital funds for equipment procurement will be managed by the Project Director in liaison with Company X.

Risk strategy

Have the CSFs changed since stage 1 completion? (As linked to the prioritized project objectives and the critical path through the project risks)
The CSFs for this sub-project have been defined. The vision of success is actually a CSF for the overall project path of success.

Have all project risks been defined and analysed? (What will stop the achievement of success?)
Some significant risks have been identified that are related to time constraints and potential cGMP issues. A full risk assessment was conducted but some technical risks will need a separate risk review as the conceptual design progresses.

What mitigation plans are being put into place?
The business change plan should address some risks and others will be managed through more specific risk reviews such as scope risk assessment, schedule risk assessment and cGMP audits.

What contingency plans are being reviewed?
There are contingency plans in place to support the detailed design and engineering and then operation of the facility in terms of providing additional cGMP expertise.

Project organization

Who is the Project Manager? John Edwards
Has a project organization for all resources been defined?
A task team organization has been developed with representative from all engineering functions within Company B, as well as product development chemists from Company X and a regulatory manager from Company Y. Towards the end of the project it is anticipated that representative from Company A would need to be involved to support handover for detailed engineering and design.

Contract strategy

Has a strategy for use of external suppliers been defined? (The reasons why an external supplier would need to be used for any part of the scope)
There will be no external suppliers required for this sub-project; however the team have been asked to give advice to the Project Director who is responsible for the development of the contract plan.

Is there a process for using an external supplier? (For example selection criteria, contractual arrangements, performance management)
Any external suppliers will be selected and managed in accordance with Company X procedures suitably advised by Company B where cGMP issues arise.

(continued)

Table 6-1 (Continued)

PDP – Product ABC facility – conceptual and front end design					
Project Director:	Trish Roberts	**Revision:**	3	**Date:**	*Confidential*
Sponsor:	*Confidential*	**Status:**	Approved	**Copy no.:**	1

Project control strategy

Is the control strategy defined?

An appropriate control strategy is in place and is based around the development of a set of interim and final deliverables of an appropriate quality:

- *Cost control strategy* – The project has a man-hour plan for the expenditure and control of the engineering design fees. An earned value approach is being used to ensure effective and efficient use of resources whilst also controlling schedule (Figure 6-6). In addition, the team will advise on the rundown of capital cost contingency.
- *Schedule strategy* – The schedule has been developed into a detailed Gantt chart (logic linked and resource loaded) and this supported development of the earned value progress measurement. Apart from using schedule contingency themselves the team will also advise on the rundown of overall schedule contingency and placement of schedule buffers.
- *Change control* – Change control will work on a number of levels. Scope, schedule and man-hour budget changes all need to be evaluated with respect to time, capital cost and cGMP impact (Figure 6-7). At a key point in the project roadmap a 'design freeze' will occur and after this point changes to key principles will be challenged.
- *Action/progress management* – the progress of deliverables will use a quality matrix to ensure that they are being developed, reviewed and approved by suitably qualified personnel. This will be supplemented by the earned value tracking.
- *Reporting* – Twice a month a summary report will be issued to the Project Director. This will focus on whether the project will deliver the required critical components to allow the overall project to proceed to schedule.

Project review strategy

Is the review strategy defined? (How will performance be managed and monitored – both formal and informal reviews and those within and independent to the team?)

The Project Manager has a weekly meeting with all the task team where critical milestones are reviewed and any risks discussed. An action log methodology is to be used as well as risk and issue logs. The focus on performance management is to pre-empt problems before they exist and deal with them effectively.

Team members are expected to be proactive and to behave in a manner which aligns with the project goals. Team members can have informal 1-1's with the Project Manager on an as needs basis.

The Project Director will perform a monthly review looking into the detail of the scope development versus schedule, cost and quality targets.

The Project Director was confident that the strategy for the delivery of this sub-project was effective and in alignment with the overall goals of the project. Additional extracts from the full PDP are collated in support of the above.

Where necessary this PDP also extracted elements from the overall PDP to ensure full alignment. This is a usual tactic when dealing with a hierarchy of PDPs which interact with each other and are sub-sets of an overall strategy.

Business plan

The overall project PDP defined how the various sub-projects were to be connected (Figure 6-3). As the conceptual and front end design project was the first of these it was important that their PDP explained where all the outputs would be going.

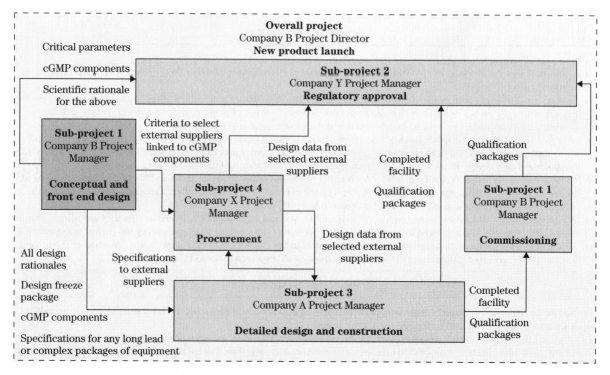

Figure 6-3 Dependency map: engineering project

The conceptual and front end design sub-project was responsible for setting the foundations of the overall project success (Figure 6-4).

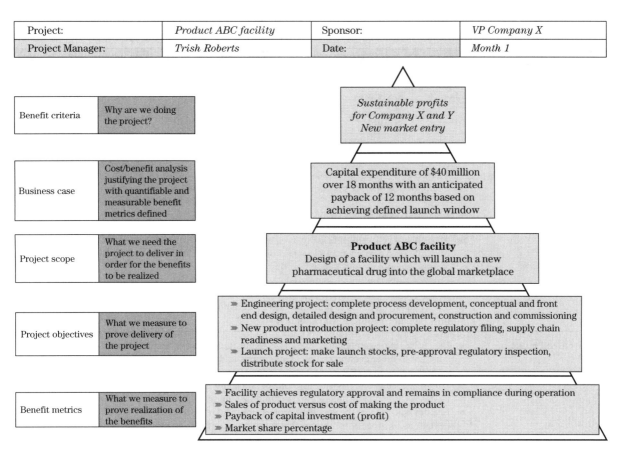

| Project: | *Product ABC facility* | Sponsor: | *VP Company X* |
| Project Manager: | *Trish Roberts* | Date: | *Month 1* |

Figure 6-4 Benefits Hierarchy: engineering project

The way the facility was designed would impact the way that it would be procured, installed, constructed and tested. It would build in compliance to regulatory needs yet still consider the overall goal of the project – a successful and timely product launch.

This project was so fundamental to the overall success that the Project Director wanted each task team member to understand the business case and why their role was so important. As a result of this the task team highlighted a number of business change issues and developed a proposed business change plan to support overall project success (Table 6-2).

Table 6-2 Business change plan: engineering project

Business change plan – Product ABC facility					
Project Director:	Trish Roberts	**Revision:**	1	**Date:**	*Confidential*
Sponsor:	*Confidential*	**Status:**	Proposed	**Copy no.:**	1

Executive summary
As a result of this project a new facility will be available to Company X. This will be the first facility which X will have had to operate in compliance with FDA regulations.

Starting point	▸ Executive, management and operations teams run existing general chemical manufacturing operations. ▸ All staff can generally access any area and can choose any route to do so. ▸ All staff try to correct errors as a part of their role and these are not generally reported – the culture is about getting it right first time. ▸ Processes are continually improved and 'tweaked' as the research chemists and operations staff learn more about the process – the focus is on cost effectiveness. ▸ The company policy is to minimize costs where possible and all staff are trained to 'switch off' company resources when not in use.
Desired end point	▸ Executive, management and operations teams must run one pharmaceutical operation adjacent to other general chemical operations and newer biopharmaceutical operations, in compliance with cGMP. ▸ The ABC facility will have access control and a gowning policy for specific areas within the facility – these will be adhered to (to minimize the potential for cross-contamination). ▸ Any deviations from the specified process must be reported and investigated. Staff needs see deviations as process conditions that **must** be reported, not merely corrected. ▸ Research chemists and operations staff understand the concept of critical parameters and qualification and know that change for any reason must be controlled and cannot be done without prior approval. ▸ The ABC facility will have areas which can **never** be switched off if the environment is to stay under control (For example the HVAC must operate continuously).
Change approach	▸ Explain key issues to the executive team: facility access, environmental controls, change control, and their role in getting staff to sustain cGMP compliance. ▸ Integrate the key research chemist and the proposed facility ABC operations manager within the conceptual and front end design team. ▸ Once the design is completed hold a GMP orientation session with the full research team (who run the pilot plan) and the operations team. Explain key design decisions in terms of maintaining cGMP compliance. ▸ Use the operations team as a part of the commissioning team so that they are trained in deviation management and process control.
Anticipated issues	▸ This represents a change in culture for all levels in the organization. ▸ It is a concern that even though the technical staff may understand the cGMP impact of changes requested by management they may not feel able to challenge them.

The Project Director reviewed this business change plan against other business change issues highlighted during the development of the overall PDP and collated one integrated business change plan.

Set-up plan

The administration of this particular sub-project was important in bringing most parts of the overall project together for the first time:

➤ Company X wanted to understand the technology transfer issues and processes as the new product and process were translated into a facility design. They also wanted to influence equipment and layout selection as these were closely linked to overall capital cost.
➤ Company Y wanted to ensure that the overall regulatory timeline was being considered.
➤ Company A wanted to influence the design and also ensure that they understood the design concepts sufficiently for detailed design to be progressed in alignment.

This first part of the project lasted 12 weeks with key stage gate decision and check points (Figure 6-5).

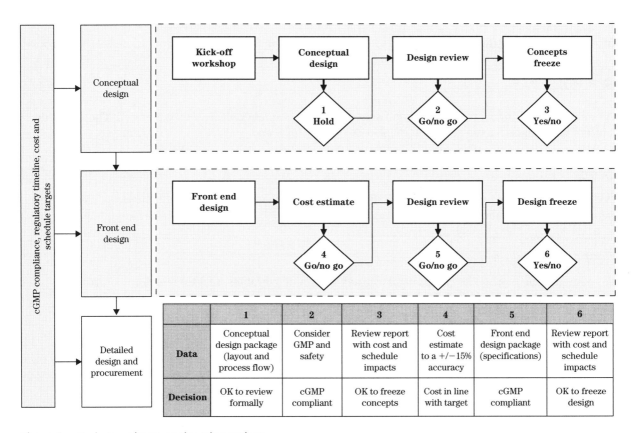

Figure 6-5 Project roadmap: engineering project

A quality matrix was used to define the quantity, quality and functionality of the scope at each stage. The primary goal of the conceptual design was to define the main process and equipment flows so that key benefits targets could be checked at stage gate 3:

➤ Facility ABC capacity to align with current market forecasts.
➤ Design is constructible within schedule targets whilst being cGMP compliant.

The main goal of the front end design was to specify equipment and facility so that compliance was effectively built in. This would minimize the risks of the detailed design being in non-compliance. In addition benefit targets were checked at stage gate 6:

➤ Facility ABC operational cycle time aligns with capacity needs.
➤ Facility capital and operational costs in line with targets.
➤ Design is constructible within schedule targets whilst being cGMP compliant.
➤ Equipment can be procured within schedule targets whilst being cGMP compliant.

Control plan

The control plan for this sub-project was fairly complex, detailed and based around the following tools:

Schedule

A complex logic linked, resource loaded schedule with schedule buffers was generated. The schedule buffers were placed in the schedule to protect stage gates 3 and 6.

Scope

A quality matrix was developed to track all deliverables with critical dates highlighted as linked to the schedule and schedule buffers.

Cost and schedule

An earned value methodology was used to track design deliverables, and therefore to control schedule and man-hour fees (Figure 6-6).

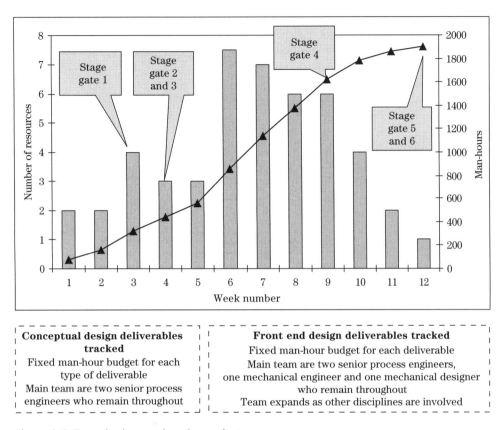

Conceptual design deliverables tracked
Fixed man-hour budget for each type of deliverable
Main team are two senior process engineers who remain throughout

Front end design deliverables tracked
Fixed man-hour budget for each deliverable
Main team are two senior process engineers, one mechanical engineer and one mechanical designer who remain throughout
Team expands as other disciplines are involved

Figure 6-6 Earned value: engineering project

Figure 6-6 demonstrates how the Project Manager planned to track the progress of various elements in one methodology. Each week the plan is to review:

➤ *Deliverables* – What tangible outputs have been delivered, at what quality, by whom and taking how many man-hours (earned value).

➤ *Earned value* – How does the earned value (in man-hours) compare with the plan and actual man-hours spent.

➤ *Schedule adherence* – Are man-hours being spent at the appropriate rate? If we are delivering the outputs with the correct number of hours but have not got sufficient resources then schedule will be impacted.

➤ *Cost adherence* – Are man-hours being spent efficiently? If we are delivering the required outputs for fewer man-hours then we are on schedule and efficient (cost savings).

➤ *Quality adherence* – Are the deliverables meeting quality standards?

Change

Change management processes were developed to control the scope against the frozen design and to manage potential impacts to cost and schedule (Figure 6-7).

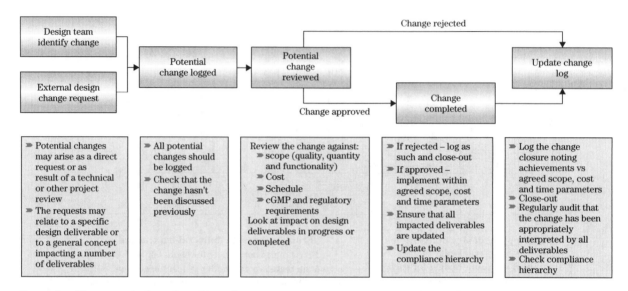

Figure 6-7 Change control: engineering project

Compliance

The aim of the strategy to control compliance was to build it in to all activities from the start of the sub-project and extend this philosophy into all other sub-projects.

- *Selection criteria for vendors and sub-contractors* – A set of criteria were developed to support the maintenance of cGMP compliance as equipment and packages progressed through detailed design, procurement, installation and testing.
- *Compliance hierarchy* – The design team used this tool as a baseline from which to understand and communicate compliance risks (Figure 6-8). For example, highlighting where the product contacts equipment and therefore where materials of construction (MOC) is an critical GMP design criteria.

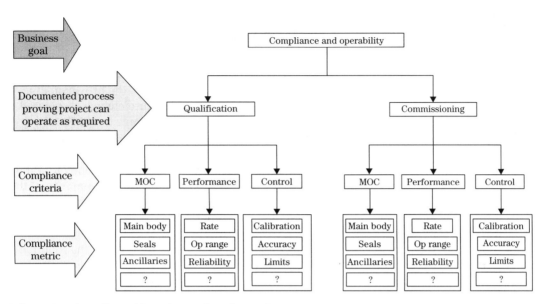

Figure 6-8 Compliance hierarchy: engineering project

The compliance hierarchy helps to design key concepts from a cGMP perspective by looking at materials of construction (MOC), operating ranges (Op range), reliability, accuracy and precision. In order to protect the patient, the equipment and process cannot introduce any contaminant and must operate within specified limits.

Conclusions

The project progressed in line with the PDP and delivered the required business benefit: support of the launch of a new drug in compliance with all required regulations. In addition, the provision of a sustainability plan, with external audit checks, supported the ongoing culture change associated with the manufacture of a pharmaceutical product and therefore sustain compliance to cGMP.

At various stages in the project issues did occur and these were usually related to the interpretation of scope quality and functionality by equipment suppliers and sub-contractors not used to working within the pharmaceutical environment. However these were resolved through implementation of contingency plans at an early enough stage so that risk impact was eliminated. The majority of these involved the mobilization of additional resources in the UK based team where the majority of GMP expertise was held. The man-hour contingency funds were prepared on this basis and therefore this strategy did not represent an engineering fees overspend.

The conceptual and front end design sub-project PDP was fundamental in setting the foundation for good project management practices. Following the success of this sub-project the Project Director was able to request some common tools be used to ensure that other sub-projects, and therefore the overall project, remained in control:

- **Earned value methodology** – using a process of assigning value to procurement packages and construction contracts so that cost and schedule adherence could be tracked simultaneously.
- **GMP audits** – using a risk based approach to check completed detailed design and installation against cGMP categories.
- **Visual controls** – using a notice board at each location to give all team members an overview of current status of the design or site activities.
- **Change control** – using a standard process and document to request, define, review and approve change ensuring that all impacts (cost, schedule, cGMP) had been fully considered.

In addition Company A was also able to share some of its detailed design tools and experiences of local suppliers and contractors. In the end more of these were used than originally planned due to the efforts from Company A and B in explaining and then checking what cGMP meant in the context of this project.

Company X and Y have gone on to work together on other product licensing deals and Company X now has a bigger pharmaceutical and biopharmaceutical portfolio which has global reach.

Lessons learnt

- Developing a real planning culture within a Project Team is often difficult when management are pushing for delivery, however it is crucial in maintaining alignment as delivery occurs.

- Planning can help bring a diverse team together and to quickly establish ground rules for project delivery.

- Virtual team working can 'work' if you spend the time on building a shared goal and on agreeing roles and responsibilities.

7

Case Study Two: strategic consultancy assignment

This case study demonstrates how the project delivery planning philosophy can be used on less traditional and less tangible projects. The project used is actually a consultancy assignment supporting the achievement of a key strategic goal for a services business.

Situation

A services business had recently completed a strategic review using both internal and external data collection. The review highlighted that although the company's strategic intent was to maintain a 'balanced' business approach, it was not achieving this.

The selected model to define, manage and measure 'balance' was the Kaplan and Norton Balanced Scorecard methodology (Figure 7-1).

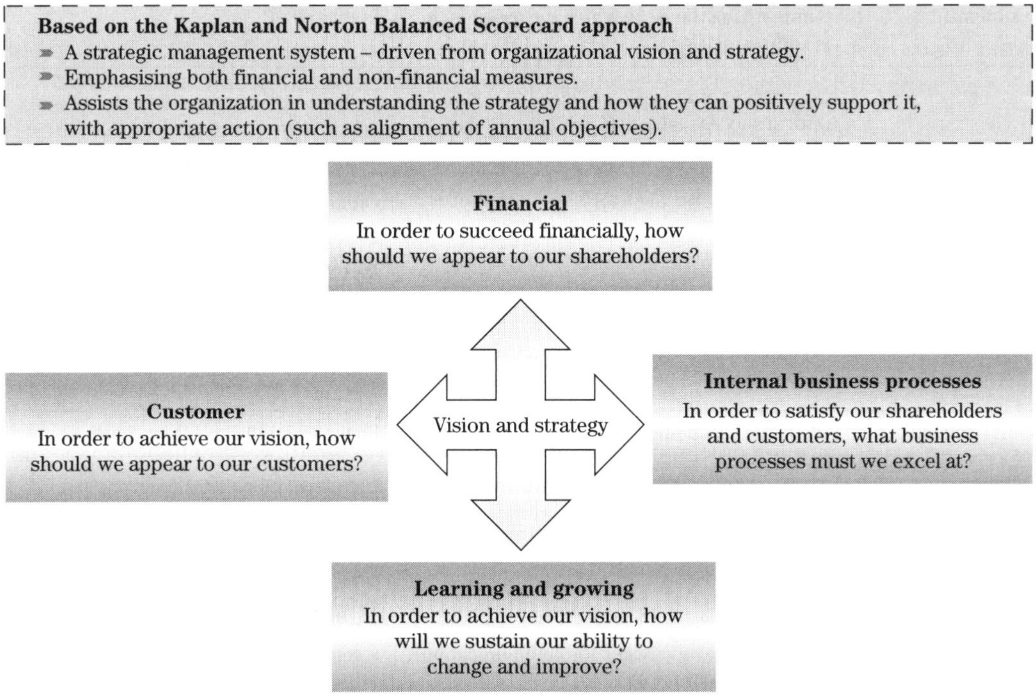

Based on the Kaplan and Norton Balanced Scorecard approach
- A strategic management system – driven from organizational vision and strategy.
- Emphasising both financial and non-financial measures.
- Assists the organization in understanding the strategy and how they can positively support it, with appropriate action (such as alignment of annual objectives).

Financial
In order to succeed financially, how should we appear to our shareholders?

Customer
In order to achieve our vision, how should we appear to our customers?

Vision and strategy

Internal business processes
In order to satisfy our shareholders and customers, what business processes must we excel at?

Learning and growing
In order to achieve our vision, how will we sustain our ability to change and improve?

Figure 7-1 The Balanced Scorecard methodology

Customer feedback in particular was 'average' and did not give the company any real understanding of how they were adding value to their customers. It also did not support them in assessing if their service portfolio was appropriate, needed expanding, contracting or developing.

The company was due to commence a major service improvement initiative aimed primarily at reducing service delivery costs; however, they were wary about commencing this with such poor customer intelligence.

A consultancy assignment was therefore approved, with the goal of providing a methodology to assess 'value to customers' and then testing it to determine if the organization was in fact, adding value. This would also support selection of appropriate service portfolio changes which would not adversely impact customers and indeed which could potentially support improved customer satisfaction.

The assignment was given to one of the senior change managers within the company who enlisted the support of an external consultant.

The external consultant took the role of Project Manager although the Assignment Lead was the change manager. The first activity progressed was the development of an assignment delivery plan (equivalent to a PDP) which would support the small assignment team in delivering the required output within 3 months (Table 7-1). Key to this activity was the development of a clearer understanding of the problem and therefore the full vision of success.

Table 7-1 Value metrics assignment: Project delivery plan (PDP) contents

Area of delivery plan	What this is	Location
Assignment charter	The sponsor approved objective and time-scale for delivery	In main PDP
Problem statement	Summarizing the issue which the assignment is addressing ('*why*')	In main PDP
Assignment path of success	Description of the vision of success and the critical success factors (CSFs) required to achieve it	In main PDP
Assignment organization	Showing '*who*' is involved	Organization chart – in main PDP Dependency chart – in main PDP ***RACI Chart*** is a separate detailed document showing specific assignment responsibilities
Assignment scope	'*What*' activities will be performed and what deliverables will be developed	Hierarchy of scorecards – in main PDP Table of CSFs – in main PDP
Assignment roadmap	'*How*' the assignment will be delivered	Phase 1 and Phase 2 flow charts – in main PDP
Project schedule (milestone chart)	An overview of '*when*' the key activities will occur	High-level milestones incorporated into the assignment roadmap – in main PDP ***Project progress plan*** is a separate detailed document showing detailed activities to be performed each week
Communications plan	Identifying key messages for communication key stakeholder groups	A separate ***stakeholder map*** has been generated The ***Communications Plan*** is integrated into the ***assignment action log***

Assignment Delivery Plan

The Assignment Delivery Plan (equivalent to a PDP) was developed by the Project Manager, reviewed by the Change Manager (assignment lead) and the Project Team members, and approved by the sponsor. Sponsor approval focused on key elements of the plan, notably the final deliverable and the method by which this would be achieved.

Executive summary

Problem statement

There is a concern that the organization does not adequately demonstrate 'delivery of customer value' to its customers. This is seen by:

- Poor customer feedback – both formal survey results and informal ad hoc feedback.
- Internal challenge – the requirement to articulate the level of customer value delivered next year (financial and non-financial) is now an organizational objective.

The definition of 'value metrics' is seen as a challenge in itself:

- How can the benefits of customers engaging with the organization be clearly identified and then communicated (to those internal and external to the organization)?

The assignment

This is a 3-month assignment to develop and test a value metrics model that would demonstrate to the organization and its customers that it had delivered an increase in value to customers. That value would be both financial and non-financial and would be delivered against robust value metrics targets.

In delivering the above output the assignment would address the defined problem statement and would therefore deliver:

- A system to measure and analyse customer value.
- A method for ongoing collection of value metrics as a part of business as usual (BAU).
- A scorecard to demonstrate value added by the service organization to specific customers.

Assignment objective

To provide and populate a value metrics model that would demonstrate (to the service business and its customers) that the services delivered had increased value within the customer organization. That value would be both financial and non-financial and would be delivered against robust value metrics targets.

This delivery plan

This assignment delivery plan is a collation of documents which describe '*how*' the assignment is to be delivered:

- It is based around the delivery of the requested assignment objective.
- It details '*why*', '*what*', '*when*', '*who*' and '*how*'.
- It provides a baseline from which to assess if the assignment delivery is '*in control*'.

Essentially this plan is an internal team document used 'live' throughout the 3-month assignment (Phase 1) to check that progress is in line with business needs. It is based on an assignment charter, previously approved by the business management team, and has been approved by the sponsor.

Business plan

The assignment is linked to the achievement of an organizational objective and therefore has the support of the entire business management team. The member of this team responsible for customer relations is responsible for the delivery of the specific organizational objective: delivery of a specified level of value to a specific set of customers (both financial and non-financial).

Project charter

Before project launch, and based on the approved business case, an assignment charter (Table 7-2) was developed by the Project Manager and Assignment Lead.

Table 7-2 Value metrics Project Charter

Planning Toolkit – **Project Charter**		
Project: Value metrics	**Assignment Lead:**	Phil Davies
Date: Week 1	**Page:**	1 of 2
Project description	**Project delivery**	
Sponsor Fred Smith – Customer Relations Manager	**Project Team** Phil Davies, Fiona Lord (Project Manager), Simon Jones (finance) and Sarah Dibble (change projects)	
Customer The customer of this assignment is the user and eventual owner of final deliverable – this has yet to be decided. However it is likely to be one of the management team or their delegated resources	**Additional resources** Additional resource to collect customer data and also service data will be required. The management team have authority over the required resources	
Project aim A sustainable process to demonstrate value delivered to customers to aid strategic decision making and support improved customer satisfaction	**Critical success factors (CSFs)** ➤ Understand customer value (their perspective) and translation into value metrics ➤ Accurate and appropriate data ➤ Robust assignment planning and delivery	
Project objectives ➤ Provide a methodology and model to measure value delivered to the customers of the services business ➤ Measure value delivered to a specific set of customers for a specific set of services delivered by the business	**Organizational dependencies** ➤ Customer relations BAU and project work ➤ Customer survey results (good starting point) ➤ Current BAU metrics ➤ Current service change/improvement projects	

(continued)

Table 7-2 (Continued)

Planning Toolkit – **Project Charter**			
Project:	*Value metrics*	**Assignment Lead:**	*Phil Davies*
Date:	*Week 1*	**Page:**	2 of 2
Project description		**Project delivery**	

Project description	**Project delivery**
Benefits ➤ Improve customer satisfaction ➤ Improve strategic decision making regarding service delivery (more customer facing)	**Risk profile** *Opportunities* – to find out more about our customers and to react positively to this to increase customer satisfaction *Threats* – lack of data, lack of ownership, lack of use and sustainability if the answers don't support current ideas for change, potential to decrease customer satisfaction, value may not be identified
Final deliverable A business process to collect, measure and communicate customer value	
Interim deliverables ➤ A method to select customers and services ➤ A collation of value metrics, defined and analysed (spreadsheet format) ➤ A data collection plan (what, why and how) ➤ A draft scorecard (output of draft business process)	**Critical milestones vs deliverables** ➤ Select customers – week 4 ➤ Service selection – week 6 ➤ Draft scorecard and business process – week 10 ➤ Issue Revision 0 (Rev 0) scorecard to customers – week 12
	Project delivery approach Select some customers, select some services and identify 5–10 value metrics that demonstrate value has been added

The charter was reviewed by the Project Team during launch and then approved by the sponsor. It was also communicated to the management team. This gave the team a robust baseline from which to plan and deliver the project.

Business change issues

As a part of understanding how the final deliverable would 'fit' with BAU, the team developed an understanding of all performance measures as a hierarchy of scorecards (Figure 7-2). This was used extensively during sponsor and stakeholder discussions to add clarity to what this assignment would, and would not, deliver.

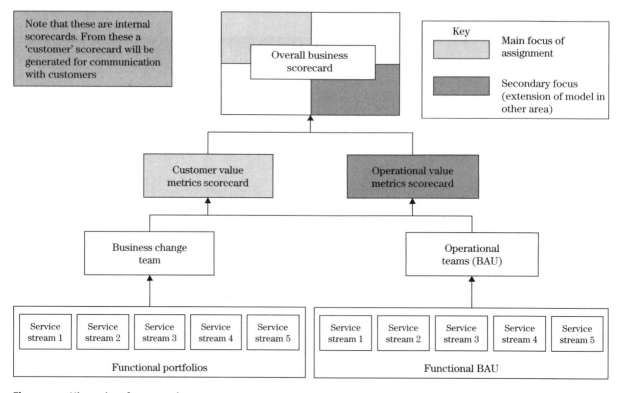

Figure 7-2 Hierarchy of scorecards

The hierarchy demonstrated that the final deliverable from this assignment would be an addition to current business performance measures systems. The difference being that the value metrics scorecard would be customer facing rather than internally service facing. Potential business change issues were identified:

- Changes to data collection processes within the service streams both in terms of '*what*' is collected and '*how*'.
- Changes to the way that service streams made decisions on what service criteria to provide, (or not), to a customer.
- Changes to demand management processes which needed to be segmented by customer **not** service stream.

However, the business was already undergoing a transformation to make it more customer facing, so a clear understanding of what else was happening in the services business and the impact on the assignment was necessary. The dependency map (Figure 7-3) was used during stakeholder discussions to clarify who was taking the lead in certain areas of the business transformation and how information should flow around the business to maintain an aligned approach.

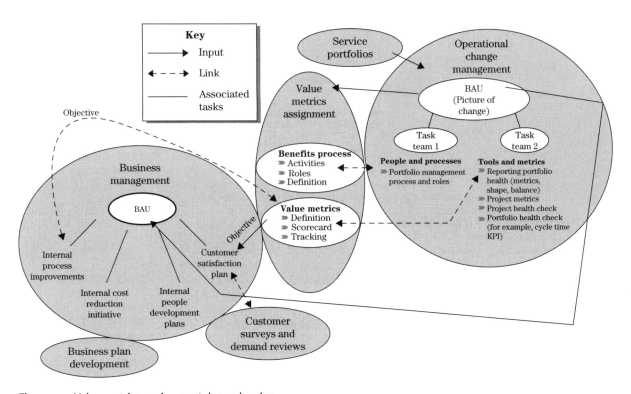

Figure 7-3 Value metrics assignment dependencies

Benefits definition

The key to benefit definition was a further development of the project critical success factors (CSFs) to review the impact these would have on the project outcome. As a result the original three CSFs were expanded and clarified and a more detailed vision of success developed (Figure 7-4).

CSF 1 **Customer value definition and delivery** Ensure that it is clear *what* customers value and that we are doing activities which will deliver this. Select customer, metrics and projects/BAU and then *deliver* accordingly	CSF 2 **Scorecard design, delivery and use** Development of a process to track a specified set of value metrics and to report on those metrics, ensuring that the process and output does not duplicate current activities. Design for usability and ease of interpretation	CSF 3 **Robust data and data collection** Ensure that there is accurate data readily available and accessible which can be analysed to produce evidence of value generation. Data collection must be lean and not duplicate current data collection processes
CSF 4 **Robust and lean links to BAU** Effective links to all parts of BAU. Strategies to cope when this work is impacted by an activity or it impacts another activity	CSF 5 **Effective stakeholder management** The identification and effective management of all stakeholders (internal and customers): using appropriate strategies to gain buy-in and understanding	CSF 6 **Assignment delivery plan developed and delivered** The development of a robust plan to deliver all assignment CSFs in order that the vision of success is successfully achieved by the end of next year and the assignment is delivered 'in control'

Vision of success
- Development of a sustainable value metrics process linked to our business processes, for example, decision making and strategic planning.
- Demonstrated delivery of value (to selected customers) next year
  - Value delivery as per baseline
  - Accurate and appropriate scorecard easily communicated and understood
- Improve selected customers view of our organization as 'value adding' to them and their business
- The management team buy-in to this process as value adding for them and for the services business as a whole

Figure 7-4 Value metrics path of CSF

This enabled the team to expand on the benefits of this assignment, giving the business a baseline and target from which to measure the achievement of the vision of success (Table 7-3).

The vision of success also highlighted the key sustainability metrics which would need to be checked following completion and handover of the assignment:

- An assigned owner for the new scorecard generating process – incorporating data collection and analysis as a part of service stream BAU, collation and verification of data.
- Support of the process from the service stream managers (allowing resources and agreeing changes to data and measures for their service streams).
- Use of the scorecard in communicating with customers.
- Use of the scorecard in making strategic decisions regarding the future of any one service within a service stream.
- Changes in other business processes: demand management, service efficiency measurement, customer management.

Due to the difficulty in measuring some of the benefits metrics and also the timeline required for changes to be seen, the sustainability checks are crucial in determining if the assignment has been successful.

Table 7-3 Value metrics benefits specification

Project management Toolkit – Benefits Specification Table					
Project:	Value metrics		**Date:**	Week 2	
Potential benefit	**Benefit metric**	**Benefit metric baseline**	**Accountability**	**Benefit metric target**	**Area of activity**
What the project will enable the business to deliver	*Characteristic to be measured*	*Current level of performance*	*Person accountable for delivery of the benefit to target*	*Required performance to achieve overall benefits*	*The project scope that will enable this benefit to be delivered*
Improve customer satisfaction	Customer satisfaction rating (annual review)	5.5 out of a maximum score of 10	Customer Relations Manager	6 by next year improving to 7.5 in 2 years	New business process for making service related decisions
	Customer satisfaction rating (at point of service delivery)	7 out of a maximum score of 10	Customer Relations Manager	7 by next year improving to 8 in 2 years	Communication of value delivered Improve perceptions as well as reality
Add increased value to customers	Financial value	Not measured currently	Service Stream Managers	$5 million by year end	Measurement of value already delivered
	Non-financial value (use benefits scoring)	Not measured currently	Service Stream Managers	Value benefit score of 10	Recommended changes from scorecard Rev 0
Improve services strategic decision making	Management team strategic performance (use alignment rating)	Not measured currently	Head of Business Services	Alignment rating of 95%	New business processes for collecting and analyzing data and for making service related decisions

The easiest change to measure will be any change in the business customer culture: how the service streams and management teams think of, and interact with, customers.

Set-up plan

Due to the fast-track nature of this strategic assignment the effective set up of the project and co-ordination of organizational resources was a high priority.

Assignment organization

A review of the knowledge necessary to complete the assignment was conducted and additional resources were allocated to the Assignment Team (Figure 7-5).

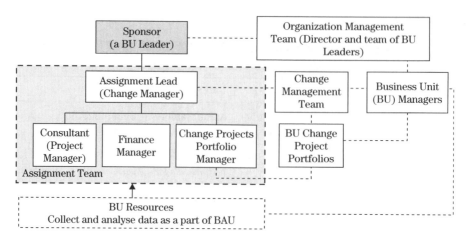

Figure 7-5 Value metrics assignment organization

A detailed RACI Chart was developed (Table 7-7, page 269) based on the proposed pace and scale of the assignment and this was developed by the team during the assignment launch event (Figure 7-6). A launch event was necessary to begin the 'real work' as a team and get agreement that there would be weekly working sessions to move the project forward at the pace required to meet the goals.

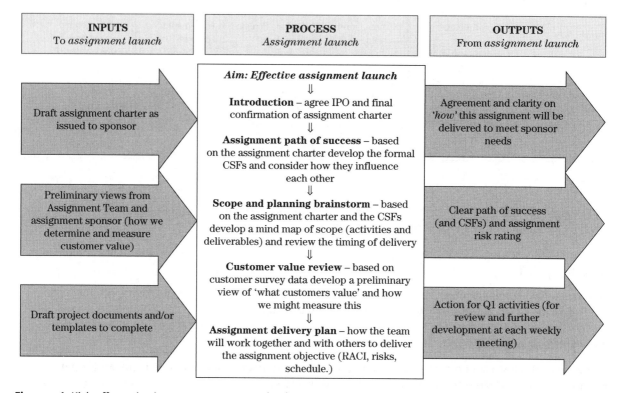

Figure 7-6 Kick-off meeting input–process–output (IPO)

Assignment roadmap

Strategic assignments have no fixed roadmap. Each time an assignment commences the Project Manager has to determine the journey which needs to be taken in order to achieve the assignment goal. Within this assignment, the journey was separated into two distinct phases:

➤ Phase 1: A 3-month period to develop and populate the value metrics scorecard by the Assignment Team, with growing interaction with service streams as the period progresses (Figure 7-7).
➤ Phase 2: A 9-month period for the business to use and test the scorecard process having access to the Assignment Team on a reducing basis over the period (Figure 7-8).

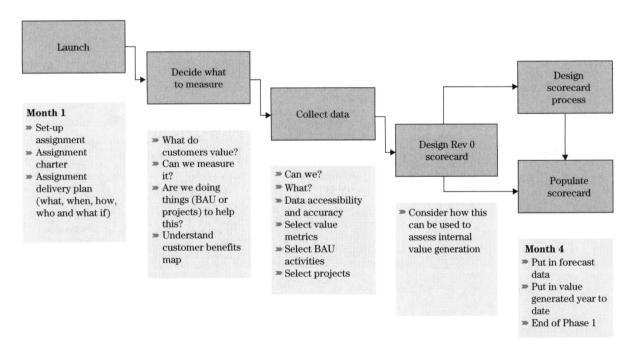

Figure 7-7 Value metrics assignment roadmap – Phase 1

The majority of the project delivery plan (PDP) is concerned with the delivery of Phase 1 of the assignment although the principles and processes set-up are valid for Phase 2 (with the exception of the weekly meetings which will be monthly in Phase 2).

Within Phase 1 there are key decision points, but these are all within the control of the Assignment Lead:

➤ Which customers to focus on?
➤ Which service streams to review?
➤ What measures to be used?
➤ What data to be collected?

All these decision points are on the assignment critical path as data from one supports the next decision. At this planning stage the exact acceptance criteria for each decision has not been determined. These are key interim deliverables and are therefore also on the critical path.

The main delivery principle for this assignment is that the team will meet each week to progress activities and review progress and performance.

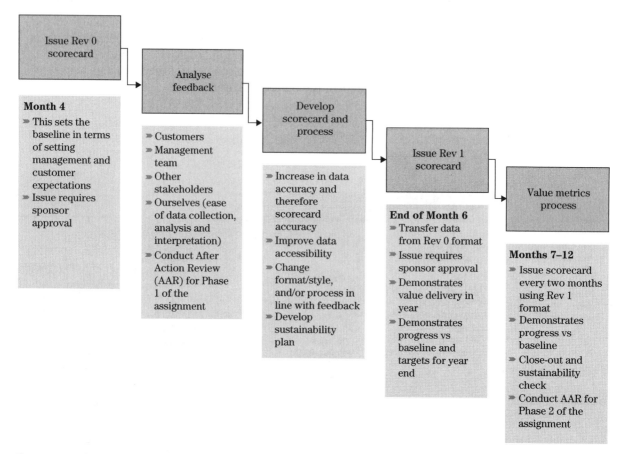

Figure 7-8 Value metrics assignment roadmap – Phase 2

Towards the end of Phase 1 of the assignment a set-up and control plan will be needed for Phase 2. These have been excluded from this PDP at this time so that the team can focus on the Phase 1 fast-track targets.

Assignment scope

A scope brainstorm against the identified CSFs clarified additional activities and interim deliverables (Table 7-4). The majority of these are to be developed during the Phase 1 weekly meeting.

Table 7-4 Value metrics table of CSFs

CSF	Deliverable	CSF	Deliverable

1	**Customer value definition and delivery**
	➤ Customer map
	➤ Customer selection strategy
	➤ Customer benefits maps
	➤ Value matrix
	  ▷ Customer value vs service activity – project and BAU
	➤ Service selection strategy and tool
	  ▷ Incorporating value metrics scoring
		2	**Scorecard design, delivery and use**
			➤ Value matrix tool (spreadsheet format)
			➤ Value metrics definition
			➤ Value metrics scorecard(s)
			  ▷ Customer scorecards (selected)
			  ▷ Service scorecards (selected)
			➤ Value metrics business process
			➤ Value metrics report
			  ▷ Interpretation of the scorecard and recommendations for the organization

3	**Robust data and data collection** ➤ Data collection plan ➤ Data owners map ➤ Data collection and analysis business process	4	**Robust and lean links to BAU** ➤ Dependency map ➤ Impact analysis assessment ➤ Sustainability plan

5	**Effective stakeholder management**
	➤ Stakeholder map
	➤ Stakeholder management plan
	➤ Communications plan
	  ▷ Assignment progress communications
	  ▷ Communicating the assignment deliverables
	  ▷ How to talk to customers in an aligned way regarding all other work
		6	**Assignment delivery plan developed and delivered**
			➤ Assignment charter
			➤ Assignment delivery plan
			➤ Detailed 'live' documents/tools
			  ▷ RACI Chart, risk rating, weekly action log and meeting outputs
			➤ Progress reports
			➤ AAR and close-out report

Control plan

Resource plan

The resources have been agreed based on having 3 days/week of the Project Manager's time, and 1 day/week of the three other Assignment Team members' time.

The data collection resources cannot be determined at this stage. They will be determined as a part of developing the data collection plan, an interim deliverable.

Progress plan

The schedule for Phase 1 has been developed in a grid format (Table 7-8, pages 270 and 271) due to the pace required. This is based on critical path analysis in terms of dependencies between activities, interim and final deliverables. The assignment roadmap for Phase 2 is considered sufficiently detailed at this stage and no further schedule development has been completed.

During Phase 1 agreed actions will also be noted and tracked through use of an action log (Table 7-9, page 272) which will be reviewed and updated at the weekly meeting.

Risk assessment

The risk profile has been assessed against the CSFs and the current risk rating is *red*. The scope and assignment approach have incorporated mitigation actions so that the risk rating will reduce to *amber* by the end of Phase 1 (month 3), and moving towards *green* as Phase 2 progresses (Table 7-5).

Table 7-5 Value metrics summary risk assessment

CSF	Mitigating actions	Status	Plan by end month 3	Plan by end month 6
1. Customer value definition and delivery	Until we can test the value matrix with customers this risk is unlikely to reduce	Red	Amber	Amber
2. Scorecard design, delivery and use	We need to identify a clear 'owner' to start to mitigate some of these risks – discussed the potential for 'finance' to own due to credibility and link to ratification of data	Red	Amber	Green
3. Robust data and data collection	Risks are unlikely to change until we are into the data collection – at the moment we are only pulling together the plan (value matrix)	Amber	Amber	Green
4. Robust and lean links to BAU	The links are clearly being built but until we can show the completed draft matrix we cannot fully test them	Red	Amber	Amber
5. Effective stakeholder management	We need to use the communications plan to engage with the key stakeholders – expect to be able to do more when we have the draft value matrix	Red	Amber	Amber
6. Assignment delivery plan developed and delivered	There are areas outside our control which are still risks to the delivery of this assignment	Red	Amber	Amber

Review plan

The performance of the assignment, and the Assignment Team, will be reviewed both formally and informally (Table 7-6).

Table 7-6 Assignment review strategy

Planning Toolkit – Review Matrix			
Project:	*Value metrics*	**Assignment Lead:**	*Phil Davies*
Date:	*Week 2*	**Page:**	1 of 1
	Internal	**External**	
Formal	➤ Weekly team meetings (action log, risk and progress plan reviews) ➤ Weekly meeting minutes	➤ Monthly progress reports (Table 7-10, page 273) ➤ Monthly 1-1s with sponsor ➤ Get an agenda for 1 or 2 of the business management team meetings during Phase 1	
Informal	➤ Assignment Team member ad hoc 1-1s and Teleconferences (TCs)	➤ 1-1s with stakeholders ➤ As a part of Assignment Team member BAU role – liaison with linked business processes and service stream managers	

Appendices

Tables 7-7 to 7-10 are included in this section.

Table 7-7 RACI Chart

Project Management Toolkit – RACI Chart

Project: *Value metrics* — **Date:** *Week 1*

R = do activity R* = complete during weekly meeting A = accountable (able to approve) C = consult A = accountable (able to approve) C* = require resource I = inform

Activity ↓	Assignment Lead – Phil	Project Manager – Fiona	Finance – Simon	Change projects – Sarah	Sponsor	Business team	Customers
1. General assignment project management	A	R	C	C	I	–	–
2. Assignment delivery plan	A	R	C	C	C (approve)	–	–
3. Assignment progress plan	A	R	C	C	I	–	–
4. Assignment risk assessment	AR*	R*	R*	R*	I	–	–
5. Assignment aim achieved (as per charter)	R	C	C	C	A	C*	–
6. Stakeholder map	AR*	R*	R*	R*	C	–	–
7. Dependency map	AR*	R*	R*	R*	–	–	–
8. Sponsor communications	AR	C	C	C	I	–	–
9. General assignment communications	A	R	C	C	I	I	–
10. Progress reports	A	R	C	C	I	–	–
11. Link to customer plans and data	I	C	I	AR	–	–	C*
12. Link to finance business processes	I	I	AR	I	–	–	–
13. Data collection plan	C	AR	C	C	I	I	–
14. Data collection	C	AR	C	C	I	C*	–
15. Value metrics spreadsheet/model	C	AR	C	C	–	–	–
16. Scorecard design and process design	AR*	R*	R*	R*	I	C*	–
17. Interpret/report value metrics scorecard	AR*	R*	R*	R*	I	I	I
18. Assignment close out report	A	R	C	C	C	I	–
19. Assignment AAR	AR*	R*	R*	R*	C	I	–

Table 7-8 Progress plan: value metrics assignment Phase 1

The following weekly plan will be used to assess progress and will be reviewed at the end of each weekly meeting attended by therefore not specifically mentioned.

						Progress Value
	Week 1	**Week 2**	**Week 3**	**Week 4**	**Week 5**	**Week 6**
Main theme of weekly meeting	Assignment launch and set-up	Planning	Learning about our customers Risk review	Measures	Measures Risk review	Data collection planning
External communications	None	Sponsor 1-1	Business review meeting	Customer review meeting	Progress report 1	Sponsor 1-1
Activities	Vision, CSFs and scope brainstorm Ground rules and roles Stakeholder mapping	Assignment planning How it 'fits' with other work	Review customer data Customer mapping	Measures mapping Customer benefits mapping	Measures mapping Kano analysis	Design of the value metrics model Review current businesss processes
Deliverables	Assignment charter and roadmap Draft RACI Standard meeting agenda	Draft PDP Deliverables list Dependency map Risk Rev 0	PDP Risk Rev 1	Customer selection strategy	Risk Rev 2	Data collection plan Value metrics model Rev 0
Critical milestones	Kick-off	PDP reviewed Customer survey data obtained	PDP approved	Select customers	Define customer value measures	Select service activities
Meeting location	Site A, XB5	Site B, A1	Site A, XB5	Site B, A1	Site A, XB5	Site B, A1
Meeting attendees	Full team	Simon out	Full team	Full team	Simon out	Full team

the Assignment Team (Phil, Fiona, Simon and Sarah). The Action Log and Progress Plan will be reviewed every week and are

Plan – Rev x, date y
metrics assignment

Week 7	Week 8	Week 9	Week 10	Week 11	Week 12
Data collection	Data collection Scorecard design	Data collection Scorecard design	Scorecard process design Populate scorecard	Populate scorecard	Phase 1 completion review
Data owner actions	Data owner actions	Progress report 2	Business review meeting	Sponsor 1-1	Progress report 3
Data collection and analysis Work with data owners	Data collection and analysis Work with data owners	Data analysis Work with data owners	Convert data into metrics Validate data Process mapping	Review collated metrics and interpret key messages	Final review of scorecard and gain approval to issue
Risk Rev 3	Value metrics model Rev 1	Risk Rev 4	Business process to generate and use scorecard	Risk Rev 5 Value metrics model Rev 2 Draft scorecard	Progress Plan Phase 2 Rev 0 scorecard
None	Selected value metrics	All data collected Scorecard designed	None	None	Issue Rev 0 scorecard to customers Business process 'live'
Sarah office	Site B, A1	Site B, CK 1	Site A, XB5	Site B, A1	Site A, XB5
Only Fiona and Sarah	Full team	Full team data owners	Simon out	Full team	Full team

Table 7-9 Action log template

This is a rolling list from previous meetings – those grayed out were completed since the last meeting. Those still 'live' remain on the list with current status noted. Those now deleted were grayed out in the previous revision.

	Action Log – Rev x, Date y Value metrics assignment			
Action number	Action	Date entered	Action by (person and date)	Current status
1				
2				
3				
4				
5				
6				
7				
8				
9				

Table 7-10 Report template

The following template will be used to issue succinct monthly reports to the sponsor and business management team.

Progress Report No. x – monthly Value metrics assignment

Assignment Lead	Phil Davies	Sponsor	Fred Smith	Assignment objective
				To provide and populate a value metrics model that would demonstrate (to the services business and its customers) that the services delivered had increased value to within the customer business. That value would be both financial and non-financial and would be delivered against robust value metrics targets

Monthly progress

Activity review	Critical milestone review	Risk review
Introductory comments: *<insert text>*	The following are the critical milestones for the first phase of the assignment:	The risk profile for the assignment is reviewed regularly and mitigating activities tracked:

Activity	Relevance to this assignment	Critical milestone	Date planned	Status	CSF	Mitigating activity updates
1. *<insert activity progressed in month>*	*<insert how this will help achieve the objective>*	Customer selection	Week 4	*<insert progress>*	1 – Customer value definition and delivery	*<insert status of mitigation plans>*
2. *<insert activity progressed in month>*	*<insert how this will help achieve the objective>*	Service activity selection	Week 6	*<insert progress>*	2 – Scorecard design, delivery and use	*<insert status of mitigation plans>*
3. *<insert activity progressed in month>*	*<insert how this will help achieve the objective>*	Scorecard design and process Rev 0	Week 10	*<insert progress>*	3 – Robust data and data collection	*<insert status of mitigation plans>*
4. *<insert activity progressed in month>*	*<insert how this will help achieve the objective>*	Issue Rev 0 scorecord	Week 12	*<insert progress>*	4 – Robust and lean links to BAU	*<insert status of mitigation plans>*
General progress comments: *<insert text>*		General comments on milestones: *<insert text>*			5 – Effective stakeholder management	*<insert status of mitigation plans>*
					6 – Assignment delivery plan developed and delivered	*<insert status of mitigation plans>*

Summary

Summary comments: *<insert text>*

Conclusions

The team were able to maintain progress at the rate planned and meet all Phase 1 critical milestones. The outputs of Phase 1 were able to give the management team a deeper insight into some of the low customer satisfaction issues they had seen in recent months. The Kano analysis (Figure 7-9) was a key deliverable in understanding that although the business was delivering some excellent services to its customers, some were unhappy with the 'basics'.

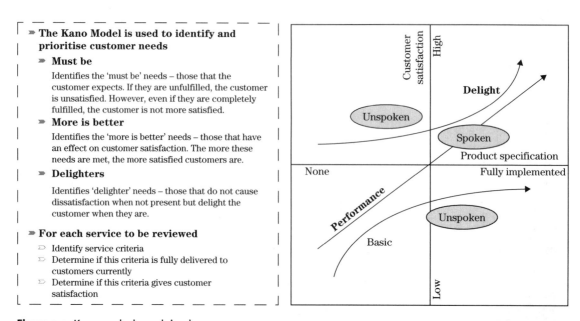

The Kano Model is used to identify and prioritise customer needs

Must be

Identifies the 'must be' needs – those that the customer expects. If they are unfulfilled, the customer is unsatisfied. However, even if they are completely fulfilled, the customer is not more satisfied.

More is better

Identifies the 'more is better' needs – those that have an effect on customer satisfaction. The more these needs are met, the more satisfied customers are.

Delighters

Identifies 'delighter' needs – those that do not cause dissatisfaction when not present but delight the customer when they are.

For each service to be reviewed

- Identify service criteria
- Determine if this criteria is fully delivered to customers currently
- Determine if this criteria gives customer satisfaction

Figure 7-9 Kano analysis explained

As a result of such analyses the scorecard was designed to highlight:

- How improving the basic service was delivering additional value to customers.
- That key service criteria were improving as requested by customers (in line with their business needs).
- Improvements in customer productivity as a result of a more appropriate service being delivered (based on customer feedback).

The assignment did deliver the required tools and processes which enabled the business to take a more customer oriented view of their service business performance. Sustainability was effectively 'locked in' by changing the business process, which linked customer demand with service response (data on both customer value and service cost was considered during strategic decision making).

Ultimately the assignment achieved its benefits targets whilst also meeting a significant yet unstated target of changing the culture of the services business: customer facing as opposed to its previous internal view.

Lessons learnt

➤ Even the most 'fuzzy' projects need planning. Typically strategic assignments don't receive an appropriate level of real project planning and suffer as a result, occasionally delivering nothing.

➤ A project which uses only part-time resources (including the Project Manager) needs *more* thorough planning in terms of deliverable and activity accountabilities, as most work is done 'off-line' when the team are not together.

8

Case Study Three: organizational change programme

This case study has been chosen from an organizational change programme recently and successfully delivered. It demonstrates how the project delivery plan (PDP) principles, tools and the template can be used to plan a programme as opposed to one distinct project.

Situation

Due to recent restructuring and relocation of a company's technical support service, major organizational change programme was required. This programme needed to ensure that all aspects of the technical support service would continue to operate both during and after the restructuring and relocation. The service was a part of a healthcare provider's supply chain and the changes were needed to support the overall increase in the quality of healthcare provided to patients.

The manager with responsibility for Technical Services was becoming concerned that the level and pace of change required was going to have an adverse impact on his department's ability to provide the level of service needed to support patient healthcare targets (in terms of quality and quantity). He began a review of the various ongoing changes and concluded that they all had some form of link to each other and needed to be more 'connected' (Table 8-1).

Table 8-1 Technical Services Review Programme definition

Area of change	Current management mode	'Fit' within a Technical Services Review Programme
1. Building a new facility for Technical Services	Capital engineering project	YES – a major project within its own right but impacts business operations
2. Getting equipment for the new facility	Funding is in the capital project; delivery is BAU	YES – requires a review of the supply chain needs (capacity and technology)
3. Getting the new facility ready to operate	All elements identified but not being formally planned	YES – this could be divided into testing and approvals
4. Moving existing people and equipment into new facility	Not being planned at present	YES – this could impact the supply chain and needs to be well planned

(continued)

Table 8-1 (Continued)

Area of change	Current management mode	'Fit' within a Technical Services Review Programme
5. Organizational design (how the management and operations teams are structured)	Not being planned at present but a recognition that this needs to occur	YES – this needs to be looked at in the context of the overall supply chain
6. Demand management (linked to increased demand and/or service levels)	Considered a part of BAU	YES – so many activities will impact the supply chain that this needs to be considered and planned
7. Improving all current business processes and/or requirement for new ones	Considered a part of BAU	YES – this needs to be bounded so that resources can be appropriately allocated and the changes planned in alignment with all others

The senior executive agreed that this was a sensible way to proceed and so a programme was formally defined and a programme PDP developed. Due to the complexity an external Programme Manager was employed to initially plan and then deliver the programme.

Project Delivery Plan (PDP)

The PDP is presented as a complete document in the template format with the exception that some detailed appendices have been removed.

PDP – Technical Services Review Programme

Technical Services Review Programme

Project Delivery Plan (PDP)

Project	Technical Services Review Programme
Customer	Healthcare Company A
Date	January year 1
Revision	02
Author	Kath Plummer
Distribution	Technical Services Director Technical Services management team New Build Project Manager

PDP Approval

	Project role	Company	Name	Signature	Date
Author	Programme Manager	Fix Consultancy Ltd	Kath Plummer	Kath Plummer	January year 1
Reviewer	Quality Manager	Healthcare Company A	Keith Jones	Keith Jones	January year 1
Reviewer	Production Manager	Healthcare Company A	Heather MacDonald	Heather MacDonald	January year 1
Approver	Technical Services Director	Healthcare Company A	Mark Smith	Mark Smith	January year 1

Project Delivery Plan (PDP) (Continued)

PDP – Technical Services Review Programme

Technical Services Review Programme

Project Delivery Plan (PDP) – Table of Contents

Technical Services Review – Project Delivery Plan – Revision 02   *Page 1 of 36*

Project Delivery Plan (PDP) (Continued)

PDP – Technical Services Review Programme – Executive Summary

1. Executive summary

1.1 The programme

This programme relates to the current and future activities of the Technical Services department within Healthcare Company A. It is made up of a large central service unit and two smaller units that deliver some niche services to very specific parts of the organization. The units share safety, quality assurance and quality control services.

The main elements of this programme are being implemented to support the overall organizational strategy, which will see all technical services fully relocated within the next 3 years. Although this programme is only concerned with the relocation of one of the Technical Services supply chains (its central service unit) the impact of dividing the function into two separate locations needs to be considered.

Throughout the programme the goal is to assure the compliance and supply of services before, during and after the move, so as to maintain patient safety and achieve regulatory approval, where applicable.

Technical Services are also seeing this period as an opportunity to review their business and to make fundamental changes which align with their vision, the core of which is 'improved patient safety'.

Within the programme there are seven Critical Success Factors (CSFs) which need to be delivered if Technical Services is to achieve its objectives (Figure 8-1).

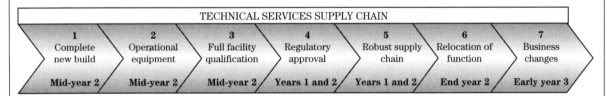

Figure 8-1 Critical path of success – Technical Services Review Programme

1.2 Programme gap analysis

One aim of this PDP is to highlight gaps in the current ad hoc programme set-up. This has mainly been completed using a structured risk assessment process following development of the programme scope; schedule and organization. Two major issues of concern are as follows:

Resource capacity gaps

- All the projects require significant time from the Technical Services management team, which is *not* available due to the 'day job', that cannot be neglected without impacting patient safety.
- A combination of internal moves within the department, internal secondments within the organization, and use of external consultants and contract (temporary) resources should be reviewed in order to balance the needs of all the projects against the day-to-day needs of the Technical Services department.

Project Delivery Plan (PDP) (Continued)

PDP – Technical Services Review Programme – Executive Summary

Resource capability gaps

➡ This is a major programme and the organization will require specific skills in terms of project management, project planning and risk assessment if the projects are to proceed to plan.

➡ There is also a need for specialist skills in validation and regulatory areas, which although already within the team, may not be available due to resource capacity gaps.

➡ The need for business change management capability may also be partially within the team, however external expertise could also be needed.

1.3 The programme delivery plan (PDP)

This programme delivery plan (PDP) has been specifically prepared for the Technical Services Review Programme.

It ensures that strategies specific to the needs of the Technical Services Review Programme are developed, implemented, communicated and constructed around best practice project and programme management. It defines the programme purpose and objectives, organization and delivery strategies required to ensure that the outcome is successful and meets the business needs.

It is a key communication tool for all in Technical Services, (all members of the department, the Project Team, the sponsor and all customer/user groups).

The PDP will be updated at regular intervals (as detailed in the PDP revision strategy). This is required to ensure that as strategies are developed and refined, they are aligned in their achievement of the programme objectives and business benefits.

The Technical Services Review Programme is made up of seven distinct pieces of work, six of which are clearly projects. A hierarchy of delivery plans is therefore an appropriate way to confirm alignment (Figure 8-2).

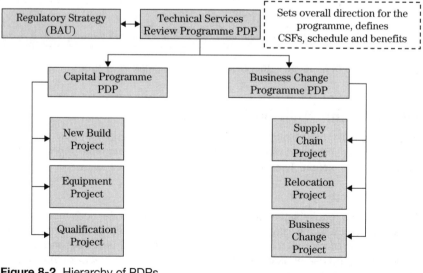

Figure 8-2 Hierarchy of PDPs

Project Delivery Plan (PDP) (Continued)

PDP – Technical Services Review Programme – Business Plan

2. Business plan

2.1 Business change management

Introduction

This entire programme is about changing all aspects of Technical Services. It is complex and there are many projects and business as usual (BAU) activities with interdependences and linkages. As a result of this project:

➤ The majority of Technical Services will be relocated.
➤ The way the department operates will change (organization, business processes, ways of working (WoW), use of technology, facilities, etc.).

The whole programme is about taking an opportunity (the physical relocation) to move closer to the organizational vision. The development of a robust system to measure the business change is required, therefore identification and tracking of the programme benefits is needed.

Technical Services: organizational position

Ultimately, Technical Services delivers a service to a patient (Figure 8-3). This value chain shows Technical Services organizational position relative to other key steps in the delivery of services to a patient.

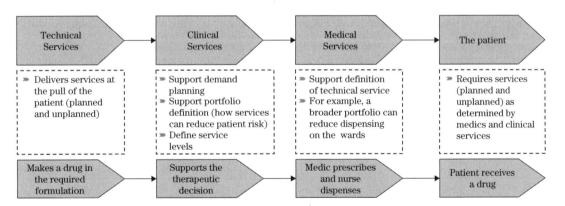

Figure 8-3 The Technical Services value chain

2.2 Business strategy

Sponsorship

Mark Smith is the Technical Services Director. He has overall responsibility for the delivery of Technical Services within the organization and is accountable for the delivery of the benefits from the successful achievement of this programme. His sponsorship of the programme is crucial to success.

Project Delivery Plan (PDP) (Continued)

PDP – Technical Services Review Programme – Business Plan

Communications strategy

Communicating the scope, objectives and progress of the programme to the organization will be critical to achieving support for its successful implementation. The key aims of the communication strategy (Table 8-2) are as follows:

- To identify the audiences and ensure that their communication needs are met.
- To build broad awareness about the programme among the management and Technical Services operations teams and among the customer group (the wider organization who use its services such as clinical and medical services).
- To communicate programme plans to all those impacted by the change, their managers and sponsors.
- To ensure that all people impacted by the change understand the change rationale and can access quality information enabling them to be motivated by the change.

Table 8-2 Communications plan

Planning Toolkit – Communications Plan

Programme: Technical Services Review	**Programme Manager:** Kath Plummer
Date: Month 1 year 1	**Page:** 1 of 2

Measures of success

The Technical Services teams are motivated and engaged in the changes included within this programme. Specific measures relate to: retention of staff during the programme, level of absenteeism, sickness, disciplinary and grievances. Apart from HR-related measures the level of participation of team members in the projects will be tracked.

Project phase	Communications objective	Communications activity

ALL phases	■ Ongoing demonstration of high-level sponsorship and support
	■ Dissemination of key messages
	■ Regular updates to all stakeholders
		■ Cascade of key messages from Mark Smith through Heather MacDonald to all in Technical Services
		■ Use appropriate communication methods for identified audiences, for example:
		  ▷ Sponsor – summary monthly reports
		  ▷ Customers – newsletters
		  ▷ Staff – team meetings and notice boards

Planning (first half of year 1 when significant planning is in progress)	■ Communicate the aims and intentions of the project to all stakeholders ■ Development of a detailed communications plan ■ As plans are developed – communicate effectively to all groups ■ Single point management of communications activities	■ Develop initial key messages (linked to the overview presentation developed as a part of the PDP development) ■ Identification of audiences and communication vehicles – understand the roles of the Technical Services management team to support effective communication ■ Preliminary stakeholder management process – identify sponsors, champions, change agents, targets, etc. and conduct a communications needs analysis across all audiences ■ Identify communications resource required (who will manage the communications activities)

(continued)

Project Delivery Plan (PDP) (Continued)

PDP – Technical Services Review Programme – Business Plan

Table 8-2 (Continued)

Planning Toolkit – Communications Plan

Programme: Technical Services Review	**Programme Manager:** Kath Plummer
Date: Month 1 year 1	**Page:** 2 of 2

Project phase	Communications objective	Communications activity
Programme delivery (pre-move)	➤ Ensure consistency of communication across all of Technical Services ➤ Support the development of commitment to delivery within each project ➤ Get all stakeholders 'excited' by the move and the associated projects	➤ Produce and launch communication (newsletters, team meetings, etc.) ➤ Develop feedback mechanisms – what is the general feeling about the move and the projects?
Programme delivery (during move – start year 2)	➤ Positive promotion of 'move' activity ➤ Demonstration and celebration of successes ➤ Demonstrate adoption of new WoW into BAU ➤ Look ahead to benefits realization for Technical Services as a whole	➤ Regular updates for stakeholders ➤ Ensure that feedback mechanisms are in place and still working
Handover and close-out (post-move – to end year 2)	➤ Positive promotion of benefits realization	➤ Regular updates for stakeholders ➤ Ensure that feedback mechanisms are in place and still working

2.3 Business benefits management

Business case

The main business case is that which Healthcare company A has determined and which has driven various building projects, including the new build associated with Technical Services. The relocation of Technical Services is a part of an overall organizational strategy and releases valuable space for use at the patient end of the value chain.

Within the overall strategy it is recognized that the Technical Services 'business' is fundamental to the organization and senior executives have approved a longer-term vision which also represents the vision for this programme (Figure 8-4). This programme should take the business closer to the vision even though it is not intended to get there within the 2-year programme time-scale. This is the 'business' that the function wants to be in.

Project Delivery Plan (PDP) (Continued)

PDP – Technical Services Review Programme – Business Plan

Vision
- To provide 'ready to administer' products to patients that cannot be provided commercially (related to either cost, time, specification or patient 'risk reduction' issues)
- To be a centre of excellence for technical services
- To provide products to patients that cannot be provided elsewhere (related to patient 'risk reduction' or niche product issues)

Strategy
- Expand capacity to meet the growing demand of a broader portfolio
- Deliver service levels as required to reduce risk to patient
- Conduct R&D to support introducing new services which improve patient safety
- Develop business processes which support the Technical Services strategy (WoW, technology, organization, customer relations, employee relations, performance improvement)

Figure 8-4 Technical Service vision (5 years)

Benefits management

Technical Services has started to identify what they need to achieve their vision and what this means in terms of organizational change. A prioritized list of benefit criteria has been defined against each of the projects to (Figure 8-5) which aligns with both the formal business case and the Technical Services vision and strategy. The primary benefit remains the link to the overall goals of the Healthcare Company A strategy – this is the highest priority for the project. The remainder of the benefits are also in order of priority.

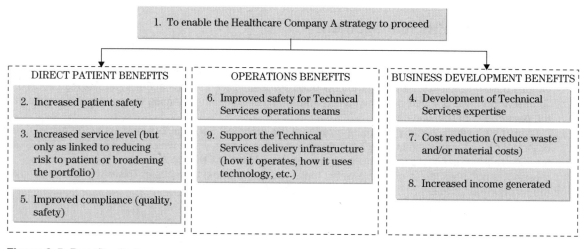

Figure 8-5 Benefit criteria map

Project Delivery Plan (PDP) (Continued)

PDP – Technical Services Review Programme – Business Plan

Based on Figure 8-5, a Benefits Specification Table has been generated (PDP Appendices, Table 8-3). This will be used to generate a Benefits Tracking Tool so that the delivery of benefits throughout the project timeline are monitored. Note that some benefits will be delivered as other projects are being implemented. The benefits specification has been used to perform a 'cross-check' against each of the projects to ensure that they are aligned to the needs of Technical Services. In essence, if any project does not support the delivery of the above benefits (in the priority stated) then it cannot be the right project.

Project completion

The Technical Services Review Programme will achieve completion when all business benefits have been realized. For each project this will imply a different point in time:

➤ Build projects are 'complete' when the physical facility is handed over with all associated documentation.
➤ Business change projects are complete when the benefits of the changes are being realized (i.e. not when the changes have been made).

Project close-out

As each specific element of the Technical Services Review Programme is completed, specific close-out activities need to be performed:

➤ Formal handover of project deliverables (the facility, the equipment and the new business processes).
➤ Financial close-out of all project budgets.
➤ Completion of a benefits realization close-out report.

These activities will be further detailed within the project PDPs as the majority of the detailed close-out activities will take place within each Project Team.

Project Delivery Plan (PDP) (Continued)

PDP – Technical Services Review Programme – Set-up Plan

3. Set-up plan

3.1 Project scope

Scope categories

There are other programmes being delivered within Healthcare Company A in parallel with this programme. Some have an impact on this programme and therefore a formal link is necessary to ensure a 2-way flow of appropriate information between programmes. In general the projects within all programmes have been categorized as either critical, build or business change (Figure 8-6).

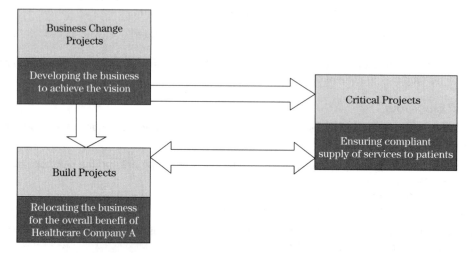

Figure 8-6 Project categories and their interactions

In addition a dependency map has been developed to show programme scope with internal and external project links (Figure 8-7).

The programme scope

The programme consists of seven major pieces of work:

1. Build Project
2. Equipment Project
3. Qualification Project
4. Regulatory Plan – this is considered BAU for the Quality Manager
5. Supply Chain Project
6. Relocation Project

Project Delivery Plan (PDP) (Continued)

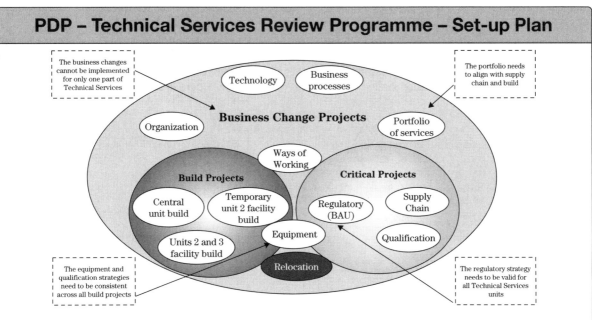

Figure 8-7 Technical Services Review Programme dependencies

7. Business Change Project
   7.1 Organization design
   7.2 Service portfolio development
   7.3 Business processes review
   7.4 WoW review
   7.5 Technology development

A one-page scope brief has been developed for each individual project (1 to 6 and 7.1 to 7.5) within the programme. They align with the programme CSF's (Figure 8-1).

Summarized and prioritized project objectives

This section defines the programme objectives. There needs to be a clear distinction between what the programme will deliver as opposed to what benefit the business will realize:

- The *benefit enabler* is what we need the project or programme to deliver in order for the benefits to be realized.
- The *benefit* is why we are doing the project.
- The *project objectives* are what we measure to prove we have delivered the project.
- The *benefit metrics* are what we measure to prove we've realized the benefits of delivering the Technical Services Review Programme.

Project priorities

It has already been stated that the highest priority benefit is the ability for this programme to enable the Healthcare Company A strategy to proceed. Therefore it would seem appropriate that the build

Project Delivery Plan (PDP) (Continued)

PDP – Technical Services Review Programme – Set-up Plan

projects which relocate the various Technical Services units are also the highest priority. However, if the compliant supply of products is not maintained this would also have an adverse impact on this benefit. Therefore the high-level prioritization of projects is as defined by Figure 8-8.

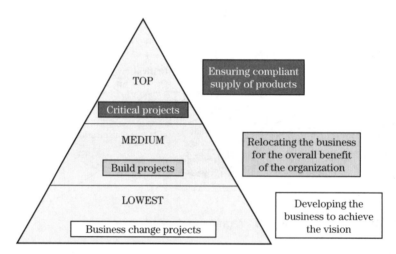

Figure 8-8 Programme priorities

Critical success factors

There is a defined critical path of success (Figure 8-1) which will deliver the programme scope and in turn enable Technical Services to achieve the benefits which support its vision.

The current progress in defining these projects is outlined in the Project Status Matrix (PDP Appendices, Table 8-4), however, this section will define the CSFs.

A CSF is some identifiable action/activity which is quantifiable and has quality attributes linked to a milestone date. It is *critical* because it has the potential to impact the overall success of the programme, portfolio or project. This means that:

- In order for Technical Services to be successful the full programme must be successfully completed (programme vision of success).
- In order for this programme to be successful all activities (projects or BAU) must be successfully completed (CSF Level 1).
- In order for each project to be successful all critical project work streams must be successfully completed (CSF Level 2).
- In order for each project work stream to be successful all critical milestones within that work stream must be successfully completed (CSF Level 3).

In this way a hierarchy of CSFs is builtup so that the lower levels align completely with the overall programme objective (PDP Appendices, Table 8-5).

Project Delivery Plan (PDP) (Continued)

PDP – Technical Services Review Programme – Set-up Plan

The vision of success for this programme is as follows:

'The successful relocation of the central Technical Services unit to the new facility without impacting the supply, to patients or the compliant status of that supply, whilst also changing the business to align with the overall Technical Services vision.'

3.2 Project organization

Overview

The overview project organization (Figure 8-9) shows the linkage of all programmes within Technical Services to the Programme Director, Mark Smith. This covers a series of programmes to fully relocate all units within Technical Services and the Programme Director is accountable for the realization of benefits which ultimately support delivery of the organizational vision.

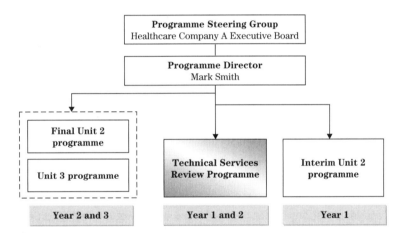

Figure 8-9 The overall project organization

Technical Services Review Programme organization

Figure 8-10 shows the organization within the Technical Services Review Programme. This is supported by a RACI Chart (A table identifying who is Responsible, Accountable, Informed and Consulted with regard to specific programme activities) and outline role descriptions.

Sponsor

The sponsor is accountable for the delivery of all the benefits from this programme (Table 8-3):

➤ His main role is to support the Programme Manager and the project management strategy through enabling appropriate funding and internal or external resourcing.

➤ He must also communicate effectively within Healthcare Company A to assure all of the ultimate success of the programme.

➤ Throughout, he should also champion the need for the programme within Technical Services so as to support the motivation of the staff to engage with the proposed changes.

Project Delivery Plan (PDP) (Continued)

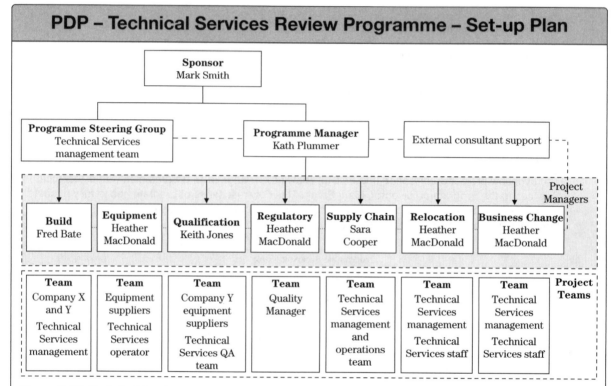

PDP – Technical Services Review Programme – Set-up Plan

Figure 8-10 Technical Services Review Programme organization

Programme Manager

The Programme Manager is responsible for the successful delivery of the Technical Services Review Programme, throughout all phases via management of the Programme Team and the PDP:

➡ Her main role is the development of an appropriate project management strategy (documented in the PDP) that assures the sponsor and Healthcare Company A of the ultimate success of the project.

➡ She then remains responsible for the delivery of this strategy (in particular for alignment at all levels) through the development of an appropriate project organization (Figure 8-10).

➡ She will manage all project dependencies through management of a team of assigned Project Managers.

➡ Throughout, she is the link with the organization via the direct reporting relationship with the sponsor, and must communicate the progress effectively at all times.

Project Manager

The Project Manager is responsible for the successful delivery of a specific project, throughout all phases via management of the Project Team and the specific PDP:

➡ His main role is the development of an appropriate PDP for the delivery of the project scope. This should detail what will be done, when, how and by whom as well as the likely costs.

Project Delivery Plan (PDP) (Continued)

PDP – Technical Services Review Programme – Set-up Plan

➤ He then remains responsible for the delivery of this PDP, tracking progress on a regular basis and reviewing risks and issues.

➤ A key role during the project will be the ongoing assessment of resource capacity and capability (linked to the issues already raised during the risk assessments).

➤ Throughout, he is the link with the Technical Services Review Programme Manager and must communicate the progress *and issues* effectively at all times.

Project Team

The Project Team is responsible for actually completing the specific activities as set out in the Programme or specific project PDPs:

➤ The Project Team will be made up of internal Technical Services resources and external companies (equipment suppliers and consultancy or contracting organizations).

➤ Project Team members need to communicate progress *and issues* with the Project Manager effectively throughout all project phases.

Overall project organization support

This is a complex business change programme requiring a variety of skills, knowledge and experience. At this stage the following additional support may be needed for the overall programme organization:

➤ Project management consultancy.

➤ Regulatory and validation consultancy.

➤ Business/organizational change consultancy.

Capability and capacity issues will be reviewed periodically and external resources utilized where appropriate.

3.3 Project roadmap

Planning project roadmap

This section is intended to describe the overall 'roadmap' which the project will take from inception to realization of the business benefits.

In such a complex programme the first step is to align the project planning so that as phased delivery commences alignment is built in (Figure 8-11). Prior to any delivery the programme and project PDPs need to be developed, reviewed for alignment and then approved.

Each project within the programme will go through the following planning stages:

➤ **Define:** This is the stage where the project CSFs are defined. Links to other projects should be clarified and the specific set of benefits which this scope will contribute to should be specified.

➤ **Plan:** Development of a detailed PDP with all the details required to ensure controlled delivery of the project CSFs.

Project Delivery Plan (PDP) (Continued)

PDP – Technical Services Review Programme – Set-up Plan

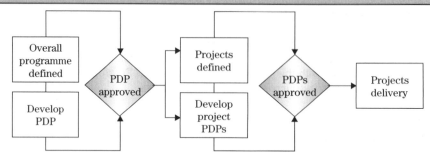

Figure 8-11 Programme roadmap – planning

Delivery project roadmap

Each project within the programme will go through the following delivery stages:

- **Implement:** For each type of project there will be a specific roadmap required. This should be clarified in each project plan. A control strategy for each project should be in place by this stage. This will link to the control strategy for the overall project (as defined within this PDP).
- **Benefits delivery:** This is the phase when the organization starts to see the benefits being delivered. The benefits specified at the start of this PDP will be tracked and the Benefits Specification Table (PDP Appendices, Table 8-3) will be converted into a Benefits Tracking Report.

Project stage gates

The overall programme and the specific projects will go through a number of distinct phases linked to approvals, funding, levels of decision making and the overall authority of the Technical Services Programme Manager and sponsor. These will be detailed within each project PDP. The generic stage gates are as follows:

- *Hold* points after design work has been completed to ensure that all appropriate approvals have been obtained and that any alignment checks have been completed across projects.
- *Hold* points before any 'go live' to ensure that all dependent activities to support sustainability have been completed or are on schedule.
- *Hold* points before any formal project closure to ensure that the project has delivered all elements of the associated and agreed CSF.

The project PDPs will schedule these stage gates individually for overall review and collation by the Programme Manager.

Project Delivery Plan (PDP) (Continued)

PDP – Technical Services Review Programme – Set-up Plan

3.4 Funding strategy and finance management

Finance sources and finance management

Within the Technical Services Review Programme there are three main sources of finance and therefore three processes for financial management:

- *Build Project budget*
  - The budget is managed by the Project Manager, Fred Bate.
  - This budget will be used for the Build Project and to partially fund the Equipment and Qualification Projects.
- *Overall Healthcare Company A Change Programme budget*
  - This budget will be managed by the executive team.
  - This budget will be used for projects associated with changes to Technical Services units 2 and 3 which are outside the scope of this PDP.
- *Healthcare Company A infrastructure revenue budget*
  - It is currently assumed that this will fund the interim Unit 2 programme, which is outside the scope of this PDP.
- *Internal Technical Services revenue budget*
  - Revenue budgets are managed by Heather MacDonald.
  - The scope which is assumed to be financed from this budget is as follows:
    - Any equipment *not* included in the main Build Project but required for the Equipment Project.
    - Internal departmental qualification resources.
    - All regulatory activities.
    - All Supply Chain Project activities.
    - All Relocation Project activities.
    - All Business Change Project activities.

Costs and funding strategy

As the project activities are developed in more detail within the each project PDP, the Programme Manager will systematically review:

- **Cost of any activity:** This will be linked to use of internal or external resources.
- **Appropriate budget allocation:** Which is the correct budget to be used for the activity and then gain approval from the budget holder.
- **Programme cost plan:** A summarized view of the costs associated with each project.

The planning stage for all projects should enable a robust cost plan to be developed so that appropriate funding can be requested and subsequently approved. Once all project cost plans have been developed they will be reviewed and collated into a summary programme cost plan.

Management fees

In delivering the Technical Services Review Programme it is noted that a number of external consultants may be required. The fees associated with these consultants are not yet explicit within the overall cost plan, however these must be estimated and a budget allocated following approval of the work. These costs should be monitored as per other items in the cost plan.

Technical Services Review – PDP – Revision 02   *Pages 9 to 16*

Project Delivery Plan (PDP) (Continued)

PDP – Technical Services Review Programme – Control Plan

4. Control plan

4.1 Contract strategy

Introduction

If the services of an external contractor or consultant are required then the following activities should be completed for each separate piece of work.

Scope of services

Define the exact scope of services required and in doing so finally review that the service cannot be provided by the organization itself. The scope should be documented.

Selection criteria

A short-list of preferred suppliers should be reviewed against specific selection criteria, as related to the scope required and the project itself. The selection should review criteria which link to specific CSFs and may be cost, quality or time related.

Contractual terms and conditions

Ensure the contract is developed in parallel with the selection process so that issues such as contract type, terms and conditions and schedule of rates, can be discussed during the selection process before order placement.

Performance review and close-out

Ensure that the contract performance is monitored throughout the project so that the contract is only closed-out upon completion of the appropriate scope at the required level of performance. If the contract has stipulated a performance test then this must be fully complied with.

Dependent on the type of contract and level of spend there will be different degrees of approval required. In general funding approval is delegated to individual Project Managers for the cost plan they are managing, with the exception of single orders over $25,000. The sponsor can approve up to $75,000 with executive board approval required for larger contracts. All management consultancy or fee-based contracts require Programme Manager review and then sponsor approval.

4.2 Risk management strategy

Introduction

The current critical path of CSFs has been defined (Figure 8-1). This will form the basis for the high-level assessment of risk and also the continuing management of risk and regular assessment regarding the potential for programme success. This is a key part of the programme control strategy. The following sections define the strategy for the management of risks and critical issues. Although only high-level programme risks and critical issues are presented in this PDP, the processes should be cascaded into the project PDPs. Although all identified risks will be managed and reported within the

Project Delivery Plan (PDP) (Continued)

PDP – Technical Services Review Programme – Control Plan

specific project, any critical issues should be reported to the Technical Services Review Programme Manager via the high-level monthly reporting system.

The risk management process

A risk is a potential issue that could impact the achievement of the programme objectives if it occurred. Once a risk is identified, it needs to be assessed:

- What is the likelihood that this risk will actually occur – low, medium or high?
- What is the potential impact if the risk did occur – low, medium or high?

Two types of plan then need to be developed:

- **Mitigation plan:** An action plan to eliminate the possibility of the risk occurring or to reduce the impact if it did occur.
- **Contingency plan:** A back-up plan to use when the risk has occurred which minimizes the impact on the project.

Central production project risk assessment

A programme risk assessment was formally conducted and a detailed table of risks was developed and analyzed. Each risk was reviewed and the following scoring system was defined:

Probability
- Low = Less than 50% likely to occur.
- Medium = 50% likely to occur.
- High = greater than 50% likely to occur.

Impact
- Low = Unlikely to significantly impact any CSF.
- Medium = Likely to significantly impact a CSF.
- High = Likely that a CSF would not be achieved.

Each risk is then plotted on a risk matrix to give a visual indicator of the category of risks: red, amber or green. Due to the number of risks identified Figure 8-12 collates the total risks in each part of the matrix.

A total of 95 risks were identified, of which 35 were found to be 'red risks'. A 'red risk' is one which is highly likely to occur and which, if it did occur, would prevent the achievement of a CSF. These types of risks have the highest priority and a mitigation plan must be developed and implemented immediately. A contingency plan should also be developed and implemented if the mitigation plan does not appear to be working:

- Each month the risks and the matrix should be reviewed so that the movement of risks can be seen.
- This movement will indicate if a specific mitigation plan is working or not and also if the contingency plan should be implemented.

Project Delivery Plan (PDP) (Continued)

PDP – Technical Services Review Programme – Control Plan

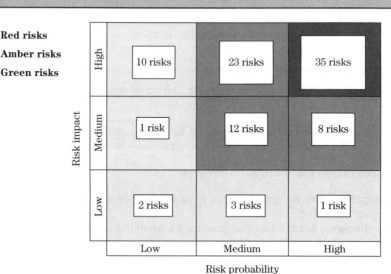

Figure 8-12 The Risk Matrix – Technical Services Review Programme

The level of red risks at the programme launch was a realistic indication of the challenge facing the team. The distribution of risks versus the individual CSFs (projects) is shown in Table 8-6. Figure 8-6 shows the impact of the level of risk in each project in the form of a 'value chain'. This allows assessment of the likelihood of success for each project, each project type and therefore the overall programme:

- Currently three of the projects are 'red' and the remaining four are 'amber'.
- One of the 'red' projects is a critical project.
- Overall this indicates that unless the risks are mitigated the likelihood of overall project success is less than 50%.

Table 8-6 also presents a reasonable plan for progressing mitigation plans based on addressing the projects in order of priority. In this way the programme should be able to be 'green' within 3–6 months by appropriately resourcing the draft mitigation plans contained in the Risk Table. The detailed Risk Table is not included in this PDP as it is a large and 'live' document which is updated at frequent periods during both planning and delivery phases.

Technical Services critical issues

A critical issue is a factor which has the potential to significantly affect the achievement of the objectives established for the project (i.e. it is a 'live' issue).

During the development of this PDP (and in particular during the risk assessment) the following critical issues became apparent. These need to be dealt with as a matter of some urgency as they have the potential to impact project success now.

Project Delivery Plan (PDP) (Continued)

PDP – Technical Services Review Programme – Control Plan

Table 8-6 Risk summary

CSF	Current Assessment				Target – Q1 year 1				Target – Q2 year 1			
	Green (G)	Amber (A)	Red (R)	Total	G	A	R	Total	G	A	R	Total
1	3	9	7	19	12	7	0	19	16	3	0	19
2	4	4	3	11	8	3	0	11	10	1	0	11
3	2	5	7	14	7	7	0	14	12	2	0	14
4	1	6	3	10	7	3	0	10	9	1	0	10
5	2	3	2	7	5	2	0	7	6	1	0	7
6	0	5	5	10	5	5	0	10	9	1	0	10
7	5	11	8	24	16	8	0	24	21	3	0	24
Total	17	43	35	95	60	35	0	95	83	12	0	95

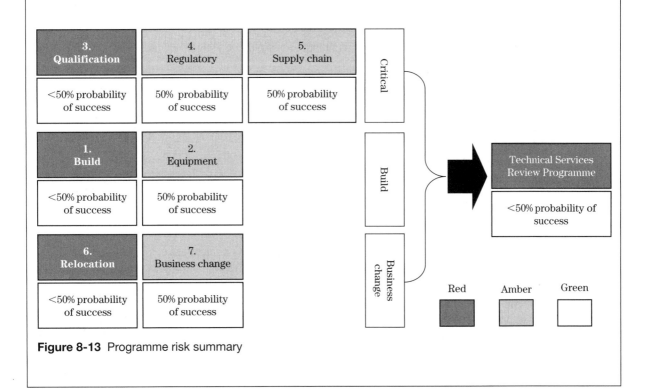

Figure 8-13 Programme risk summary

Project Delivery Plan (PDP) (Continued)

PDP – Technical Services Review Programme – Control Plan

Resource capacity gaps

➤ All the projects require significant time from the Technical Services management team which is *not* available due to the 'day job', that cannot be neglected without impacting patient safety.

➤ Build Project example – based on the Build Project programme there will need to be significant planning activity over the period April to June year 1 if the organization is to be ready to move into the building when it is completed. This is a critical issue as this activity is the highest priority milestone linked to the primary benefit of this programme. However, the current resources identified for regulatory, qualification, supply chain and business change planning already have full-time 'day jobs'.

➤ A combination of internal moves within the function, internal secondments within the organization, and use of external consultants and contract (temporary) resources should be reviewed in order to balance the resource capacity needs of all the projects against the day-to-day needs of the Technical Services function.

Resource capability gaps

➤ This is a major programme and the organization will require specific skills in terms of project management, project planning and risk assessment if the projects are to proceed to plan.

➤ There is also a need for specialist qualification and regulatory skills, which although already within the team, may not be available due to resource capacity gaps.

➤ The need for business change management capability may also be partially within the team; however external expertise could also be needed.

➤ Example – The overall integration of the projects into a coherent programme, the management of the significant number of risks and the planning and re-planning which will be needed is not a core skill for the management team.

The development of this PDP has been the mechanism for identifying the scale of the gaps identified. Due to the intensity of this programme over the next 18 months to 2 years, these two critical issues represent a major challenge to successful programme delivery over the coming months in particular. Individual projects within the programme cannot be looked at in isolation as the core teams are essentially the same – this itself represents a critical issue.

Resource will also be needed to support other organizational changes and this can only come from the Technical Services management team – representing yet another 'pull' on this team's time and energy.

4.3 Project control strategy

Introduction

Appropriate project control tools will be used to monitor progress against project objectives. Within each project plan the project controls strategy should be cascaded so that appropriate systems and procedures are in place within the Project Teams.

Programme development and milestone progress measurement

At the highest level within the programme there are milestones that are a critical part of programme success (for example, completion of the relocation by early year 2).

A high-level milestone schedule (Table 8-5) will be used as a 'live' control document for the Programme Team as linked to the CSF tracking system in the programme and the projects. Below this

Project Delivery Plan (PDP) (Continued)

PDP – Technical Services Review Programme – Control Plan

are the project schedules which are 'owned' by each Project Team. The details behind each milestone are contained within logic linked project schedules which are resource loaded. Therefore, as the schedule management is cascaded through to the Project Teams the level of detail will increase.

The high-level programme schedule has been developed initially as a dependency chart (Figure 8-14):

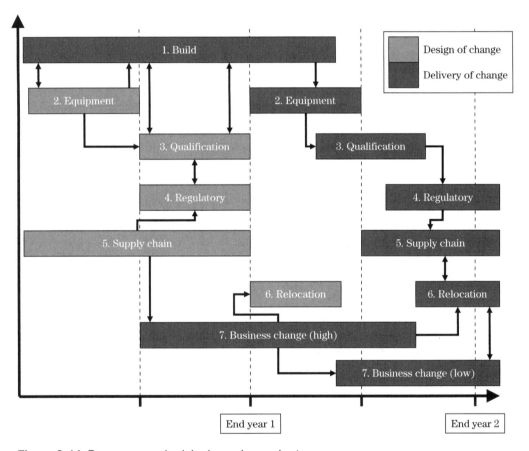

Figure 8-14 Programme schedule dependency chart

➤ This is intended to show the interactions and dependencies between the various projects but not the detail. It also shows the main links between design and delivery activities.
➤ Approximate project durations are also shown and therefore approximate project timings.
➤ The detail within each project will be developed within the project schedules themselves (to be developed by each Project Manager).

Project Delivery Plan (PDP) (Continued)

PDP – Technical Services Review Programme – Control Plan

➤ A formal Gantt Chart will be then collated from each individual project schedule and the resource loading reviewed as well as all dependencies at the activity level.

➤ The collated schedule information will then be checked for continuity in terms of: interactions between projects at activity or resource level, timing or resourcing of activities outside of the programme which may impact those within it and the ability to deliver to the timing and benefits needs of the business.

Each month the Project Managers will be reporting against agreed schedule milestones and any significant deviations reported as critical. This will ensure:

➤ Any impact on other areas of the programme can be determined.

➤ The appropriate corrective action taken at the earliest possible stage.

It is important that all projects define an appropriate progress measurement system within their plans. It is expected that a number of project control tools will be used as well as critical milestone monitoring:

➤ Critical path analysis (using logic linked schedules and associated 'S' curves).

➤ Earned value monitoring (measuring physical progress) and resource level tracking.

Cost management

The Programme Manager remains accountable for the management of the programme cost plan and has to manage this within the agreed financial processes. The Project Managers are accountable for the achievement of their specific approved cost plans and therefore they remain fully responsible for project level cost management.

Cost management begins with the development of a robust cost estimate as linked to a defined scope. Following approval of the specific Project Cost Plan, the Project Manager is accountable for the management of the expenditure against the approved budget. It is expected that the following will be tracked and reported:

➤ Cost spent versus costs approved.

➤ Forecast costs to completion.

➤ Anticipated use of cost contingency.

The Programme Manager will collate the final cost plan from all individual project cost plans and review all elements of contingency (linked to either cost, schedule or scope risk). A formal contingency run-down schedule will then be generated and used as a further control tool.

Change control

Any changes from the approved project scope briefs *must* be approved by the Programme Manager who may require sponsor approval *before* the change can proceed. Depending on the change (scope development or scope increase) funds may need to be allocated before approval.

At a project level, it is anticipated that all teams will have a change control procedure in place for all critical deliverables once they are approved (e.g. Project Cost Plan, etc.). This PDP, and all plans below it, will be subject to change control once approved.

All changes should be reviewed from both a 'hard' and 'soft' perspective, that is:

➤ Hard = schedule, scope and cost implications, etc.

➤ Soft = the 'cost' of disruption, staff well being and cultural integration, etc.

Project Delivery Plan (PDP) (Continued)

PDP – Technical Services Review Programme – Control Plan

Reporting

A summary programme report (PDP Appendices, Table 8-7) will be used to report progress against programme objectives. Within each project, the reporting strategy should be cascaded so that appropriate systems and procedures are in place. The contents of the report should be as follows:

- Key activities or achievements completed this month and planned for next month.
- Milestone or objective tracking (cost, schedule, scope).
- Benefits tracking and critical issues tracking.

The report should be issued each month commencing as soon as possible and be based on one-page monthly reports from each Project Manager.

4.4 Project review strategy

Internal reviews

The following strategy defines how the Programme Team will continually assess and manage performance and progress via reviews during the project lifecycle through all phases.

Internal informal meetings

All members of the Programme Team should maintain regular informal contact with each other. It is particularly critical in a complex programme such as this that the overall Programme Manager and the Project Managers maintain a high level of informal contact.

Internal formal meetings

A schedule of meetings will be developed confirming the following strategy:

- Monthly Technical Services Programme Steering Group meetings.
- Monthly progress review Project Team Meetings.

External reviews

The following strategy defines how the organization will continually assess and manage performance and progress via reviews during the project lifecycle through all phases:

External informal reviews

All members of the Programme Team should have access to mentors within their organization who can provide an informal forum for reviewing current performance.

External formal reviews

All elements of the programme are available for an Independent Project Review (IPR) of performance once the delivery planning phase has been completed. This type of health check can be at planned or unplanned intervals. A key milestone where an IPR can add benefit to the organization is immediately prior to project delivery – when all project strategies are in place with detailed procedures to manage and control delivery.

Technical Services Review – PDP – Revision 02   *Pages 17 to 24*

Project Delivery Plan (PDP) (Continued)

PDP – Technical Services Review Programme – Appendices

5. Appendices

5.1 Appendix I: PDP management and control

This appendix contains the proposed process for the management and control of the Technical Services Programme PDP.

Revision history

Revision number	Revisions	Issue date	Reasons for issue
0	New document – issued as a collation of the PDP checklist and PDP appendices	Month 1 year 1	For comment and review
1	Formal PDP document issued	Month 2 year 1	For review
2	Final updates	Month 2 year 1	Approved

This section details the status of the formal revision of the PDP to date.

Revision strategy

This section outlines the planned strategy for re-issue of this 'live' document. It details the proposed revisions needed and why.

Revision number	Planned revisions	Planned issue date	Reasons for issue
3	Development of areas of the PDP as appropriate to the 'live' status of the project, for example: ➡ Programme reviews ➡ Risk reviews and updates, etc.	End month 3 year 1	The PDP is a 'live' document This revision will aim to eliminate any gaps identified in Revision 2 but which did not prevent the approval
4	Development of areas of the PDP as appropriate to the 'live' status of the project	End month 6 year 1	The PDP is a 'live' document

The PDP is a dynamic document and the above intended revision strategy may be subject to change as the project progresses through its lifecycle.

Project Delivery Plan (PDP) (Continued)

PDP – Technical Services Review Programme – Appendices

Reviewer and approval strategy

This section documents who is responsible for the review and approval of the PDP.

Role	Position	Role in the PDP development process
Kath Plummer	Project Consultant	*Author* of this PDP in conjunction with the Technical Services management team (consultation and workshop process) Management of PDP development process and the PDP contents as related to the demonstration of good project management practice
Keith Jones	QA Manager	*Reviewer* of the PDP both formally and informally via the PDP workshops and the consultation process
Heather MacDonald	Production Manager	*Reviewer* of the PDP both formally and informally via the PDP workshops and the consultation process *Approval* of the PDP (from Rev 1 onwards)
Mark Smith	Technical Services Director	*Approval* of the PDP (from Rev 1 onwards)
Various	Pharmacy Technical Services	Informal review of the PDP through involvement in the PDP workshops

Distribution list

The distribution of this document is generally to the Technical Services management team. However, this PDP is *not* a 'controlled document', that is upon formal re-issue it *will* be distributed as above. However, this does not restrict it from being distributed to others within Healthcare Company A as appropriate.

5.2 Appendix II: Summary project documents

Benefits Specification Table (Table 8-3)

A completed Benefits Specification Table is included in this appendix. There are currently gaps in the table as baseline data is not available for most of the benefits.

A Benefits Tracking Table needs to be generated for a subsequent revision of the PDP but should only be done once the benefits have been completely specified (with baseline data included).

Project status matrix (Table 8-4)

This document details the current status of each project within the Technical Services Review Programme. This document can be used as a high-level status 'health check' (i.e. keep monitoring until all boxes in the table are 'ticked').

Project Delivery Plan (PDP) (Continued)

PDP – Technical Services Review Programme – Appendices

Table of CSFs (Table 8-5)

The table of CSFs should be read in conjunction with the scope briefs and the associated RACI Charts. This table only details those milestones in each project which are critical to the success of that project.

The Technical Services Review Programme report (Table 8-7)

An example of the proposed Technical Services Review Programme monthly project report is included within this appendix. This is a blank template and is intended for use by the Technical Services Programme Manager on a monthly basis. It is very high level and focuses on CSFs and risks to the achievement of CSFs.

Project Delivery Plan (PDP) (Continued)

PDP – Technical Services Review Programme – Appendices

Table 8-3 Benefits Specification Table

Project Management Toolkit – Benefits Specification Table

Potential benefit	Benefit metric	Benefit metric baseline	Responsibility	Benefit metric target	Project
(What the project will enable the business to deliver)	*(Characteristic to be measured)*	*(Current level of performance)*	*(Responsible for delivery of the benefit to target)*	*(Required performance to achieve overall benefits)*	*(What project scope will enable this benefit to be delivered)*
To enable the Healthcare Company A strategy to proceed	Relocations proceed to schedule	N/A	Fred Bate	No penalties	All projects in the programme can impact this benefit, however the key project is the Build Project
Increased patient safety	Number of new 'high risk' products produced	0	Mark Smith	1 per quarter	Supply Chain Project Business Change Project
	Decreased number of regulatory inspection audit points	4 major 10 minor		0 major <5 minor	Regulatory Plan (BAU) Qualification Project
Increased service level	Number of products in the portfolio	25	Mark Smith	40 by end year 3	Supply Chain Project Business Change Project
	Range of products in the portfolio	TBA		TBD	Supply Chain Project Business Change Project
	Increased number of 'ready to administer' products	10		20	Supply Chain Project Business Change Project
	Overall production capacity	200,000 units		600,000 units	Supply Chain Project Build Project Equipment Project Relocation Project Business Change Project

Project Delivery Plan (PDP) (Continued)

PDP – Technical Services Review Programme – Appendices

Table 8-3 (Continued)

Project Management Toolkit – Benefits Specification Table

Potential benefit	Benefit metric	Benefit metric baseline	Responsibility	Benefit metric target	Project
	Response time to clinical staff (acute)	8 hours		4 hours	Business Change Project Technical Services BAU
	Plan adherence to clinical staff (acute)	65%		85%	Business Change Project Technical Services BAU
Development of Technical Services expertise	Improved capability	Current profile TBA	Heather MacDonald	Required profile TBD	Business Change Project Succession Planning Processes to be incorporated
	Increased external reputation (increased number of referrals)	5 per quarter		15 per quarter	Business Change Project
	Introduce external courses	0 per yr	Heather MacDonald	2 per yr	Business Change Project
Improved compliance	Decreased number of regulatory inspection audit points	4 major 10 minor	Keith Jones	0 major <5 minor	Regulatory Plan (BAU) Qualification Project
	Decreased number of internal audit actions	TBA		TBD	Regulatory Plan (BAU) Qualification Project Business Change Project
Improved safety for Technical Services operations teams	Decreased number of 'safety incidents' per year	1 major 25 minor	Heather MacDonald	0 major <10 minor	Business Change Project
	Reduction in risks during operations, (manual handling, etc.)	Amber profile		Green profile	Business Change Project Supply Chain Project

Project Delivery Plan (PDP) (Continued)

PDP – Technical Services Review Programme – Appendices

Table 8-3 (Continued)

Project Management Toolkit – Benefits Specification Table

Potential benefit	Benefit metric	Benefit metric baseline	Responsibility	Benefit metric target	Project
Cost reduction	Improved production efficiency (decreased level of rejects)	15% rejects	Mark Smith	<5% rejects	Business Change Project Qualification Project
	Reduced drug wastage	TBA		50% reduction	Business Change Project
	Reduced starting material costs	TBA		25% reduction	Business Change Project
Increased income generated	Increased production units available for sale	TBA	Mark Smith	50% increase	Business Change Project
	Increased income	TBA		50% increase	Business Change Project
Support the Technical Services delivery infrastructure	Increased production units per day (linked to % utilization of available capacity)	TBA	Heather MacDonald	TBD	Business Change Project Supply Chain Project
	Increased number of working hours/week (that the unit is open)	50 hours		84 hours	Business Change Project
	Increased number of production hours (linked to % utilization of available capacity)	78%		86%	Business Change Project This is a factor related to capability and shift working – it's about making the shift patterns effective

PDP – Technical Services Review Programme – Appendices

Table 8-4 Project status matrix

Programme CSF	Scope	Strategy	Schedule	Budget	Status report	External roles	Technical services role	Priority
1. Build Project	✓ Scope brief developed from user perspective	✓ Agreed with Project Manager	✓ Draft schedule developed but major gaps relating to the programme interfaces	✓ Approved project funds	✗ Build in progress and on schedule	Sponsor Keith Jones PM – Fred Bate Main contractor – Company X Clean room contractor – Company Y	Customer contact – Heather MacDonald (set requirements and accept completed area)	High
2. Equipment Project	✓ Scope brief developed	✗ Preliminary view of categorization	✓ Draft schedule developed but requires links to Build Project	✗ Not identified- funds external to capital project	✗ Work in progress to identify and categorize	None identified but likely link to specialist equipment vendors	Sponsor and PM – Heather MacDonald Project Team – see organization chart	High
3. Qualification Project	✓ Scope brief developed	✗ Validation Master plan not yet developed	✓ Draft schedule developed	✓ Funds requested as a part of capital project	✗ Scope and external support definition in progress	Validation contractor (via organizational procedures)	Sponsor – Heather MacDonald PM – Keith Jones Project Team – see organization chart	Very high
4. Regulatory Plan (BAU)	✓ Scope brief developed	✗ Ideas formulated but not formally written down	✓ Draft schedule developed – more detail needed	✗ Not identified	✗ Discussions with regulators in progress	None identified	Heather MacDonald (as part of BAU role)	Very high
5. Supply Chain Project	✓ Scope brief developed	✗ Not developed	✓ Draft schedule developed	✓ Minor internal funds for overtime, stock buy in, etc.	✗ Capacity – Demand model being developed	None identified	Sponsor – Heather MacDonald PM – Sara Cooper Project Team – see organization chart	Very high

PDP – Technical Services Review Programme – Appendices

Table 8-4 (Continued)

Programme CSF	Scope	Strategy	Schedule	Budget	Status report	External roles	Technical services role	Priority
6. Relocation Project	✓ Scope brief developed	✗ Not developed	✓ Draft schedule developed but not complete	✗ Not identified	✗ Trying to understand the best way to approach this	None identified but likely link to specialist move company	Sponsor and PM – Heather MacDonald Project Team – see organization chart	Medium
7. Business Change Project	✓ To be developed as five separate projects	✗ Change NOW and to link to critical projects	✗ To be developed as five separate projects	✓ Implement through internal budgets	✗ Adhoc change in progress to be aligned	Potential requirement for specialist project and business change consultants	Sponsor and PM – Heather MacDonald Project Team – as per RACI Chart	High
7.1 Organization design	✓ Scope brief developed	✗ Not developed	✗ In draft outline only	✓ Internal budgets – to be reviewed	✗ Strategy workshops in progress	Potential to require external support and facilitation	Sponsor – Heather MacDonald Project Team – as per RACI Chart	TBA
7.2 Service portfolio development	✓ Scope brief developed	✗ Not developed	✗ In draft outline only	✓ Internal Budgets – to be reviewed	✗ Discussing a broader portfolio	None identified	Sponsor – Heather MacDonald Project Team – as per RACI Chart	TBA
7.3 Business processes review	✓ Scope brief developed	✗ Not developed	✗ In draft outline only	✓ Internal budgets – to be reviewed	✗ Discussing current issues	Potential to require external support and facilitation	Sponsor – Heather MacDonald Project Team – as per RACI Chart	TBA
7.4 WOW review	✓ Scope brief developed	✗ Not developed	✗ In draft outline only	✓ Internal budgets – to be reviewed	✗ Discussing extended days and shift patterns	Potential to require external support and facilitation	Sponsor – Heather MacDonald Project Team – as per RACI Chart	TBA
7.5 Technology development	✓ Scope brief developed	✗ Not developed	✗ In draft outline only	✓ Internal budgets – to be reviewed	✗ Discussing use of IT and automation	Specialist equipment vendors	Sponsor – Heather MacDonald Project Team – as per RACI Chart	TBA

Project Delivery Plan (PDP) (Continued)

PDP – Technical Services Review Programme – Appendices

Table 8-5 Table of CSFs

Project Management Toolkit – Table of CSFs

Programme: Technical Services Review	**Programme Manager:** Kath Plummer
Date: Month 1 year 1	**Page:** 1 of 3

Programme vision of success

The successful relocation of the central Technical Services unit to the new facility without impacting the supply of compliant services to patients, whilst also changing the business to align with its vision.

CSF definition

Sub-project (CSF Level 1)	Objective tracking metric (CSF Level 2)	Critical milestone (CSF Level 3)	Accountable for CSF3	Priority in project
1 – Build complete *Compliant production area delivered 'on time' and containing all production equipment suitably qualified to ensure compliant performance*	1.1 Compliant design of production areas	➤ Design review (date) ➤ DQ (Design Qualification) audit (date)	Fred Bate	Highest
	1.2 Qualified clean rooms	➤ IQ (Installation Qualification) witnessed (date) ➤ OQ (Operational Qualification) witnessed (date)	Fred Bate (contracted to Company Y)	High
	1.3 GMP compliant production areas	➤ GMP review (DQ) ➤ Final GMP review after clean down (date)	Fred Bate (contracted to Company X)	High
	1.4 Critical equipment installed and qualified	➤ IQ witnessed (date) ➤ OQ witnessed (date) – key performance criteria		High
	1.5 Production area hand-over	➤ Handover using critical features checklist (date) ➤ Compliance documentation handover (date)	Fred Bate and Heather MacDonald	High
2 – Equipment operational *All equipment, required to support business needs, is operational and compliant at the appropriate time*	2.1 Equipment identified	➤ Equipment categorization complete (date)	Heather MacDonald	Highest
	2.2 Equipment designed to meet performance needs	➤ Specifications approved (date), DQ (date)		High
	2.3 Equipment procured in a compliant manner	➤ Purchase orders approved – documentation requirements (date), DQ (date)	Heather MacDonald and Fred Bate (Build Project)	High
	2.4 Equipment installed in compliance	➤ Installation specification approved (date), IQ (date)		High
	2.5 Equipment tested in a compliant manner	➤ Commissioning test reports approved (date), OQ (date)	Keith Jones	High
	2.6 Equipment 'online'	➤ GMP status confirmed (date)		High

(continued)

Project Delivery Plan (PDP) (Continued)

PDP – Technical Services Review Programme – Appendices

Table 8-5 (Continued)

Project Management Toolkit – Table of CSFs				
Programme: Technical Services Review			**Programme Manager:** Kath Plummer	
Date: Month 1 year 1			**Page:** 2 of 3	
CSF definition				
Sub-project (CSF Level 1)	**Objective tracking metric (CSF Level 2)**	**Critical milestone (CSF Level 3)**	**Accountable for CSF3**	**Priority in project**
3 – Facility and equipment qualified *All central pharmacy processes and equipment validated in line with current regulatory requirements so as to be fully compliant*	3.1 Identify processes to be validated	➤ VMP Rev 0 (date) ➤ VMP approved (date)	Keith Jones	Highest
	3.2 Validate identified processes	➤ DQ (date), IQ (date) ➤ OQ (date), PQ (date)		High
	3.3 Identify equipment to be validated	➤ VMP Rev 0 (date) ➤ VMP approved (date)		High
	3.4 Validate identified equipment	➤ DQ (date), IQ (date) ➤ OQ (date), PQ (date)		High
	3.5 GMP procedures identified	➤ VMP Rev 0 (date) ➤ VMP approved (date)	Sara Cooper	High
	3.6 GMP SOPs written and approved	➤ SOP Plan (date) ➤ SOPs written (date) ➤ SOPs approved (date)		Medium
	3.7 Training of new GMP SOPs	➤ Training plan (date) ➤ Training complete (date)		Medium
4 – Regulatory approval achieved *All production licences approved so that the supply of products can be achieved in line with the business needs*	4.1 Identify all manufacturing licences required	➤ Regulatory plan Rev 0 (date) ➤ Regulatory plan approval (date)	Heather MacDonald	Highest
	4.2 Maintain current licences	➤ Audit operation within the limits of the current licences (date)		High
	4.3 Regulatory plan for all licences	➤ Timetable Rev 0 (date) ➤ Timetable approved (date)		High
	4.4 Development of new licences	➤ Licences developed (date) ➤ Licence internal approval (date)	Sara Cooper	Medium
	4.5 Approval for all licences	➤ New licence approval (date) ➤ Change of site approval (date)	Heather MacDonald	High

(continued)

Project Delivery Plan (PDP) (Continued)

PDP – Technical Services Review Programme – Appendices

Table 8-5 (Continued)

Project Management Toolkit – Table of CSFs

Programme: Technical Services Review	**Programme Manager:** Kath Plummer
Date: Month 1 year 1	**Page:** 3 of 3

CSF definition				
Sub project (CSF Level 1)	**Objective tracking metric (CSF Level 2)**	**Critical milestone (CSF Level 3)**	**Accountable for CSF3**	**Priority in project**
5 – Security of supply chain maintained *Technical Services supply chain managed in line with the business needs (and therefore patient requirements)*	5.1 Capacity/demand model developed	➤ Model Rev 0 (date) ➤ Model approved (date)	Sara Cooper	Highest
	5.2 Supply plan for year 1 developed	➤ Stock build up (date) ➤ Stop supply – old (date) ➤ Start supply – new (date)		High
	5.3 Acute supply maintained	➤ Supply chain audit (date)		High
	5.4 Planned supply maintained	➤ Supply chain audit (date)		Medium
6 – Relocation of Technical Services (central unit) *Technical Services relocated in line with the business needs, maintaining patient safety and employee motivation*	6.1 Move plan developed	➤ Move plan Rev 0 (date) ➤ Move plan approval (date)	Sara Cooper	Highest
	6.2 Phase 1 non-licensed	➤ Move achieved (date)		High
	6.3 Phase 2 – licensed	➤ Move achieved (date)		High
	6.4 Infrastructure moves	➤ Move achieved (date)		Medium
	6.5 People moves	➤ Move achieved (date)		Medium
	6.6 People trained	➤ Training complete (date)		Medium
	6.7 People informed	➤ Communications on time		Medium
7 – Business changes completed *Technical Services operate in alignment with the vision*	7.1 Organizational design implemented	➤ Design approved (date) ➤ Roles appointed (date)	Heather MacDonald	Highest
	7.2 Service portfolio expanded	➤ Service portfolio plan (date) ➤ Service expansion (date)		High
	7.3 Business processes are 'live'	➤ Design complete (date) ➤ Go live (date)		High
	7.4 New WoW implemented	➤ Flexible day design (date) ➤ Flexible day go live (date) ➤ Skill audit (date)	Sara Cooper	High
	7.5 Technology developments implemented	➤ Technology plan (date) ➤ Technology changes implemented		High

Project Delivery Plan (PDP) (Continued)

PDP – Technical Services Review Programme – Appendices

Table 8-7 Programme report template

Monthly Update – *<Insert Month and Year>*

Programme summary

Key activities this month	Achievements
➡ *<insert most important activities completed this month – include any relevant BAU work such as regulatory activities>*	➡ *<insert most significant milestone achievements or CSF tracking>*
Key activities next month	**Critical issues**
➡ *<insert most important activities to be completed next month – include any relevant BAU work such as regulatory activities>*	➡ *<insert risks which are becoming potential issues to the overall project>*

Projects summary

	1. Build	2. Equipment	3. Qualification
KEY ACTIVITIES THIS MONTH	➡ *<insert most important activities completed this month>*	➡ *<insert most important activities completed this month>*	➡ *<insert most important activities completed this month>*
KEY ACTIVITIES NEXT MONTH	➡ *<insert most important activities to be completed next month>*	➡ *<insert most important activities to be completed next month*	➡ *<insert most important activities to be completed next month>*
ACHIEVEMENTS	➡ *<insert milestone achievements>*	➡ *<insert milestone achievements>*	➡ *<insert milestone achievements>*
CRITICAL ISSUES	➡ *<insert movement of risks if more critical>*	➡ *<insert movement of risks if more critical>*	➡ *<insert movement of risks if more critical>*
	5. Supply Chain	**6. Relocation**	**7. Business Change**
KEY ACTIVITIES THIS MONTH	➡ *<insert most important activities completed this month>*	➡ *<insert most important activities completed this month>*	➡ *<insert most important activities completed this month>*
KEY ACTIVITIES NEXT MONTH	➡ *<insert most important activities to be completed next month>*	➡ *<insert most important activities to be completed next month>*	➡ *<insert most important activities to be completed next month>*
ACHIEVEMENTS	➡ *<insert milestone achievements>*	➡ *<insert milestone achievements>*	➡ *<insert milestone achievements>*
CRITICAL ISSUES	➡ *<insert movement of risks if becoming more critical>*	➡ *<insert movement of risks if becoming more critical>*	➡ *<insert movement of risks if becoming more critical>*

Technical Services Review – PDP – Revision 02   *Pages 25 to 36*

Conclusions

Initially there was a lot of resistance to pulling the various areas of change together into one programme. However, the support from the build Project Manager influenced many to take this proposed approach on board. In fact it gave him the forum to voice some of his concerns regarding how the facility was to be used and the lack of a coherent business change plan.

In addition the focus on business change supported a much more successful relocation of the team within Technical Services:

- The anticipated loss of staff was much lower.
- The organization was able to offer flexible working to counter the lengthier commuting time for most staff.
- The operations team started to see the benefits of the move and used it as an opportunity to engage with the business changes, getting more involved and making their opinions known about 'what doesn't work now'.

The risk assessment process proved to be a useful programme management tool, particularly as some of the anticipated red risks became a reality. Strong contingency planning by the Programme Team supplemented with further external project management, change management and qualification resources supported ultimate programme success.

At the end of the 2-year programme Technical Services was able to relocate the majority of its 'business' to the new building and managed to maintain compliant supply throughout the relocation period. The business had to adapt to running a service across two sites but this was just another factor in the design rather then an issue for the team. The benefits were realized as the department was able to offer some innovative services to the organization that reduced risks to the patient and were also saleable outside of the organization (thus delivering a new revenue stream).

Lessons learnt

- A PDP can be used to pull a programme of very diverse projects together assuming that they have some link – usually activities or benefits.

- All projects have the potential to change the business – not just those concerned with business change. A good PDP will address this.

- Not all project management planning tools are used on all projects or programmes. A good Project Manager will be selective and use or adapt a tool to meet the needs of the project or programme.

9 Appendices

Appendix 9-1 – The 'Why?' Checklist

Project Management Toolkit – 'Why?' Checklist

Project:	<insert project title>	Project Manager:	<insert name>
Date:	<insert date>	Page:	1 of 1

Sponsorship

Who is the sponsor? (The person who is accountable for the delivery of the business benefits)
<insert the name of the person who is taking this role>
Has the sponsor developed an external communication plan? (How the sponsor will communicate with all stakeholders in the business?)
<insert any comments on how the sponsor has/is communicating with the business>

Business benefits

Has a business case been developed?
<insert comments on the current status of the formally developed business case which supports the project>
Have all benefits been identified? (Why is the project being done?)
<insert comments on the progress of the articulation of the benefits of completing the project>
Who is the customer? (Identify all stakeholders in the business including the customer)
<insert comments on the completion of the stakeholder analysis>
How will benefits be tracked? (Have they been adequately defined?)
<insert comments on benefits metrics>

Business change

Will the project change the way people do business? (Will people need to work differently?)
<consider if the project will change the way that 'normal business' is conducted>
Is the business ready for this project? (Are training needs identified or other organizational changes needed?)
<consider what else is being done in other parts of the business related to the project>

Scope definition

Has the scope been defined? (What level of feasibility work has been done?)
<insert comments on the accuracy of the scope of the project>
Have the benefit enablers been defined? (Will the project enable the benefits to be delivered when the project is complete?)
<insert comments on how the scope is linked to the business benefits>
Have all alternatives been investigated? (Which may include *not* needing the project)
<insert comments on all alternatives to this project which have been considered>
Have the project success criteria been defined and prioritized?
<consider the areas of scope which the project requires to be completed in order to deliver the business benefits>

Stage One decision

Should the project be progressed further? (Is the business case robust enough for detailed planning to commence?)
<insert the decision – yes or no – with comments>

Appendix 9-2 – The 'How?' Checklist

Project Management Toolkit – 'How?' Checklist

Project:	<insert project title>	Project Manager:	<insert name>
Date:	<insert date>	Page:	1 of 3

Stage One check

Have there been any changes since Stage One completion? (Development of the business case and project kick-off may be some time apart)
<insert any changes that may impact the delivery of the project or the associated benefits>

Sponsorship

Who is the sponsor? (The person who is accountable for the delivery of the business benefits)
<confirm the name of the person who is taking this role>
Has the sponsor developed a communication plan?
<insert a comment on how the sponsor intends to communicate with project stakeholders during the project>

Benefits management

Has a benefits realization plan been developed?
<insert any data related to the schedule for delivery of the agreed benefit metrics>
How will benefits be tracked? (Have they been adequately defined?)
<insert any additional data which further articulates the specific benefit metrics and which align with work completed during Stage One>

Business change management

How will the business change issues be managed during the implementation of the project? (Are there any specific resources or organizational issues?)
<insert specific plans for the management of the business change associated with the project>
Have all project stakeholders been identified? (Review the stakeholder map from Stage One)
<attach the stakeholder analysis work that has been completed>
What is the strategy for handover of this project to the business? (Link this to the project objectives)
<insert specific plans for project handover>

Scope definition

Has the scope changed since Stage One completion? (Has further conceptual design been completed which may have altered the scope?)
<insert details of the further work which may have been conducted prior to project kick-off>
Have the project objectives been defined and prioritized? (What is the project delivering?)
<Attach a copy of the prioritized objectives>
<insert an updated list of project CSFs>

(continued)

Appendix 9-2 – The 'How?' Checklist (Continued)

Project Management Toolkit – 'How?' Checklist

Project:	*<insert project title>*	**Project Manager:**	*<insert name>*
Date:	*<insert date>*	**Page:**	2 of 3

Project type

What type of project is to be delivered? (For example engineering or business change)
<insert the type of project being delivered – note that this is a major category>
What project stages/stage gates will be used? (Key milestones for example funding approval, which might be go/no go points for the project)
<insert the project roadmap for the type of project within the organization>

Funding strategy and finance management

Has a funding strategy been defined? (How will the project be funded and when do funds need to be requested?)
<insert the funding request requirements – estimate accuracy, funding timeline, authorization process>
How will finance be managed?
<confirm that no additional reporting outside of the project control strategy is required>

Risk and issue management

Have the CSFs changed since Stage One completion? (As linked to the prioritized project objectives and the critical path through the project risks)
<insert updated critical path of success if available>
Have all project risks been defined and analysed? (What will stop the achievement of success?)
<comment on any high priority risks>
What mitigation plans are being put into place?
<attach a copy of the high priority mitigation plans>
What contingency plans are being reviewed?
<attach a copy of the high priority contingency plans>
<attach a copy of the Risk Table and Matrix>

Project organization

Who is the Project Manager?
<insert the name of the Project Manager who will be delivering the project in line with the project delivery plan>
Has a project organization for all resources been defined? (Include the Project Team and all key stakeholders)
<insert any comments on the project resource situation – capacity or capability>
<have roles and responsibilities been defined? Attach the RACI Chart and/or project organization chart>

Contract and supplier management

Has a strategy for use of external suppliers been defined? (The reasons why an external supplier would need to be used for any part of the scope)
<insert a copy of the contract plan>
Is there a process for using an external supplier? (For example selection criteria, contractual arrangements, performance management)
<confirm that procedures to manage supplier selection and performance are in place>

(continued)

Appendix 9-2 – The 'How?' Checklist (Continued)

Project Management Toolkit – 'How?' Checklist			
Project:	<insert project title>	**Project Manager:**	<insert name>
Date:	<insert date>	**Page:**	3 of 3

Project controls strategy
Is the control strategy defined? <comment on each of the following:> ➥ Cost control strategy ➥ Schedule strategy ➥ Change control ➥ Action/progress management ➥ Reporting <insert what methodologies, tools or processes will be used to ensure control?>

Project review strategy
Is the review strategy defined? (How will performance be managed and monitored – both formal and informal reviews and those within and independent to the team?) <comment on the plan for reviewing project performance during the delivery of the project>

Stage Two decision
Should the project be progressed further? (Is the project delivery strategy robust enough for project delivery to commence?) <insert the decision – yes or no – with comments on the robustness of the project delivery plan (PDP)>

Appendix 9-3 – Project Delivery Plan (PDP) Template

The following template includes guidance on the appropriate content and it should be read in conjunction with previous chapters for further explanation of individual sections.

Project Delivery Plan (PDP) Template – Title Page

<insert project title>

Project Delivery Plan (PDP)

Project	*<insert project title>*
Customer	*<insert customer organization>*
Date	*<insert date of this revision of the PDP>*
Revision	*<insert revision number>*
Author	*<insert name of Project Manager who has developed the PDP>*
Distribution	*<insert names of those who will receive a copy of the PDP>*

PDP Approval

	Project role	Company	Name	Signature	Date
Author	*Project Manager*	*<insert company or specific part of organization>*	*<insert name>*	*<author to sign master copy of PDP>*	*<author to write in date when signed>*
Reviewer	*<insert role>*	*<insert company or specific part of organization>*	*<insert name>*	*<reviewer to sign master copy of PDP>*	*<reviewer to write in date when signed>*
Approver	*sponsor*	*<insert company or specific part of organization>*	*<insert name>*	*<sponsor to sign master copy of PDP>*	*<sponsor to write in date when signed>*

<Project Title>: Project Delivery Plan – Revision <insert revision number>   *Page X of XX*

Appendix 9-3 – The PDP Template (Continued)

PDP Template – Table of Contents

<insert project title>

Project Delivery Plan (PDP) – Table of Contents

<Project Title>: Project Delivery Plan – Revision <insert revision number>   *Page X of XX*

Appendix 9-3 – The PDP Template (Continued)

PDP Template – Executive summary

1 Executive summary

The executive summary is the part of the project delivery plan (PDP) that most people will read first, if not the only part. As such, it should summarize the PDP, be able to stand alone as a logical, clear concise summary of the PDP and highlight the key issues that the reader should be aware of. It should report on the results of the project delivery planning rather than the reasoning behind them. Things on which to focus:

- The definition of the business need.
- Relationship to the strategic/business plan.
- Selected delivery methodology.
- Costs, benefits and risk profile.

There is no need to include:

- Assumptions and constraints (unless they are key).
- Background (except for perhaps one sentence).
- Analysis, reasoning or details of any form (remember that the remainder of the PDP will contain all required details).

The executive summary should be developed after the rest of the PDP has been completed and would typically contain three main sections.

1.1 The project

Write a brief paragraph on the overall goal of the project – the vision of success. Highlight any issues which came up during the planning process.

1.2 Business background

Write a brief paragraph on the reason why the project is needed and the business benefit it will deliver. Highlight any critical business change elements which are *not* included in the project but which are critical to success, the realization of the business benefits.

1.3 The PDP

Write a brief paragraph explaining the aim of the PDP so that readers understand how it will be useful for them. Bear in mind that the PDP has a wide audience and different stakeholders will use it in a variety of ways. Explain how the Project Manager will use the PDP as a part of the overall controlled delivery of the project.

<Project Title>: Project Delivery Plan – Revision <insert revision number>　　　　　　*Page <insert page number>*

Appendix 9-3 – The PDP Template (Continued)

PDP Template – Business Plan

2 Business plan

The business plan is the part of the PDP that senior stakeholders will be particularly interested in as it defines the link between the project and the business. It should clearly and succinctly summarize '*why*' the business needs the project, '*how*' the business will integrate the project into 'business as usual' (BAU) and '*what*' benefit metrics will be tracked to prove that the approved business case has been delivered successfully.

Three planning themes are dealt with in this section:

2.1 Business strategy

Go into some detail on the business context into which the project fits. In particular, identify the sponsor and the way in which both the Project Manager and sponsor will communicate to those external to the project.

Typically the following tools would be used and/or referenced at this stage:

- Sponsor Contract Planning Tool.
- Sponsor Contract.
- Sponsor–Project Manager RACI Chart.
- Project Charter.
- Communications map.
- Communications Planning Tool.

2.2 Business benefits management

This part of the PDP will define why the project is necessary and how the business will benefit from successful delivery of the project.

Typically the following tools would be used and/or referenced at this stage:

- Simple Benefits Hierarchy.
- Benefits Specification Table.
- Benefits Realization Plan.

2.3 Business change management

This part of the PDP will define how the business will change as a result of the delivery of the project and/or how the business will need to change in order to support delivery of the project.

Typically the following tools would be used and/or referenced at this stage:

- Stakeholder Management Plan.
- Stakeholder Contracting Checklist.
- Business change plan.
- Sustainability Plan.

Appendix 9-3 – The PDP Template (Continued)

PDP Template – Set-up Plan

3 Set-up plan

The set-up plan is the part of the PDP that most stakeholders will be interested in as it details the administration and launch of the project. It should clearly and succinctly summarize '*what*' the project will deliver, '*who*' is involved in that delivery, '*how*' organizational resources are accessed and '*what*' overall route the project will follow.

Four planning themes are dealt with in this section:

3.1 Project organization

This part of the PDP will define the internal organizational resources that are to be used and how they are to be organized and managed. Typically the following tools would be used and/or referenced at this stage:

- Project organization chart
- Role descriptions and/or role profiles (Project Team Role Profile) and RACI Charts
- Skills matrix and/or team selection criteria (Team Selection Matrix) and project capability profile
- Team Start-up Checklist and launch event

3.2 Project roadmap

This part of the PDP will define the route that the project will follow including the major hold or decision points as required by the organization. It will describe the project stages as linked to a specific project type as well as outline how the more generic phases of the project are to be progressed. Typically the following tools would be used and/or referenced at this stage:

- Project roadmap flow chart
- Roadmap Decision Matrix

3.3 Funding strategy

This part of the PDP will define how funding is to be initially accessed and then how it needs to be managed in order to comply with organizational governance processes. Typically the following tools would be used and/or referenced at this stage:

- Funding strategy flow chart
- Finance Strategy Checklist

3.4 Project scope

This part of the PDP will define the quality, quantity and functionality of the project objectives and deliverables. It defines what the project will actually deliver. Typically the following tools would be used and/or referenced at this stage:

- Scope Definition Checklist
- Path of critical success factors (CSFs) and Table of CSFs
- Work breakdown structure and activity plans
- Quality matrix, deliverables listing and scope risk assessment

Appendix 9-3 – The PDP Template (Continued)

PDP Template – Control Plan

4 Control plan

The control plan is the part of the PDP that will mainly be of interest to Project Team members as it defines how they will work as they deliver the project; how the project remains in control.
It should clearly and succinctly summarize '*how*' each part of the project is to be delivered in control considering cost, schedule, quality, quantity, functionality and risk. Four planning themes should be dealt with in this section:

4.1 Contract strategy

This part of the PDP will define the external resources required to deliver the project and how these are to be selected and managed. Typically the following tools would be used and/or referenced at this stage:

- Contract Plan
- Supplier Selection Matrix

4.2 Risk management strategy

This part of the PDP will define how uncertainty is to be managed both in terms of opportunities and threats. Typically the following tools would be used and/or referenced at this stage:

- Risk table, matrix and Critical Path of Risks Table
- Risk Management Strategy Checklist
- Risk checklists and decision flow charts
- SWOT and FMEA analyses
- Project Scenario Tool

4.3 Project control strategy

This part of the PDP will define how the project is to be controlled and focuses on the delivery of the specified scope within defined cost and time. It defines how change is to be managed so that the project remains able to be delivered successfully. Typically the following tools would be used and/or referenced at this stage:

- Project schedule (Gantt and PERT Charts) and cost plan
- Change control flow chart
- Resource histograms and 'S' curves, cost and schedule risk plans

4.4 Project review strategy

This part of the PDP will define how the performance of the project will be reviewed considering both internal performance management and external independent health checks. Typically the following tools would be used and/or referenced at this stage:

- Review matrix
- Project Scorecard

Appendix 9-3 – The PDP Template (Continued)

PDP Template – Appendices

5 Appendices

To ensure that the PDP is 'readable' the appendices are used to collate more detailed information which supports the main body of the document. There are usually two main sections:

5.1 PDP management and control

The PDP would normally be one of the first documents within a project that is subject to document control. This is because it is a key document which many people will use to support different activities. Therefore any changes to the PDP will need to be highlighted, reviewed, approved and then communicated to all who are using it.

Changes to the delivery strategy of a project can impact many aspects of its delivery and so changes must be analysed for impact on dependent activities.

5.2 PDP deliverables

The appendices are used to either contain or list key documents referenced in the main body of the PDP. They are usually the constituent parts of the plan which are then used to track progress throughout delivery. A typical PDP deliverables listing would include:

- Reference documentation list (referencing key documents upon which the PDP is based, such as the approved business case and other organizational reference papers)
- Project Charter
- Benefits Realization Plan
- Project Communications Planning Tool
- Table of CSFs
- Risk assessment and associated mitigation and contingency plans
- Project schedule (including critical milestones and schedule buffers)
- Project cost plan (including cash flow analysis and cost contingency plan)
- Project RACI Chart (including role descriptions as appropriate)
- Reporting schedule (including a sample report template showing agreed project KPIs to be tracked)

<Project Title>: Project Delivery Plan – Revision <insert revision number>   *Page <insert page number>*

Appendix 9-4 – The Simple Benefits Hierarchy Tool

Project Management Toolkit – Simple Benefits Hierarchy Tool

Project:	<insert project title>	Sponsor:	<insert name of assigned sponsor>
Project Manager:	<insert name of assigned Project Manager>	Date:	<insert date>

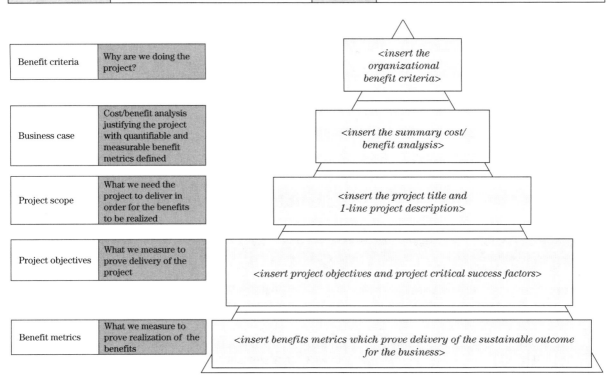

Benefit criteria	Why are we doing the project?

<insert the organizational benefit criteria>

Business case	Cost/benefit analysis justifying the project with quantifiable and measurable benefit metrics defined

<insert the summary cost/benefit analysis>

Project scope	What we need the project to deliver in order for the benefits to be realized

<insert the project title and 1-line project description>

Project objectives	What we measure to prove delivery of the project

<insert project objectives and project critical success factors>

Benefit metrics	What we measure to prove realization of the benefits

<insert benefits metrics which prove delivery of the sustainable outcome for the business>

Appendix 9-5 – The Benefits Specification Table

Project Management Toolkit – Benefits Specification Table

Project:	<insert project title>		Date:		<insert date>
Potential benefit	**Benefit metric**	**Benefit metric baseline**	**Accountability**	**Benefit metric target**	**Area of activity**
What the project will enable the business to deliver	Characteristic to be measured	Current level of performance	Person accountable for delivery of the benefit to target	Required performance to achieve overall benefits	The project scope that will enable this benefit to be delivered
<insert the benefit criteria which this benefit metric is related to>	<insert the specific benefit metric which is to be measured>	<insert the measured level of current performance, 'before' project commencement>	<insert the name of the person who is accountable for the delivery of this specific benefit>	<insert the target level of the metric at the target date>	<insert the area of the project which is specifically related to the delivery of this benefit>

Appendix 9-6 – The Stakeholder Management Plan

Project Management Toolkit – Stakeholder Management Plan

Project:	<insert project title>		Date:			<insert date>
Individual stakeholder analysis						
Stakeholder name/role	**Type**	**Current project knowledge**	**Level of Influence**	**Current level of engagement**	**Target level of engagement**	**Management of stakeholder**
<insert stakeholder's name and role in the organization>	<insert type>	<insert low, medium or high>	<insert low, medium or high>	<insert low, medium or high and a comment on the evidence to back this up>	<insert low, medium or high and a comment on the actions needed>	<insert Project Team member's name>
Summary stakeholder analysis						
<insert comments on the overall impact of the current stakeholder situation on the project>						

Appendix 9-7 – The Benefits Tracking Tool

Project Management Toolkit — Benefits Tracking Tool

Project:		*<insert project title>*		Date:		*<insert date>*
Benefit metric		**Baseline**	**Milestone 1**	**Milestone 2**	**Milestone x**	**Target**
		<insert baseline date>	*<insert date or activity>*			*<insert target date or activity>*
Metric 1 *<insert metric from benefits specification table>*	Plan	*<insert baseline data>*	*<insert planned metric level>*			*<insert target metric level>*
	Actual	*<insert baseline data>*	*<insert actual metric level>*			*<insert actual metric level>*
Metric 2	Plan					
	Actual					
Metric x	Plan					
	Actual					

Appendix 9-8 – The Sustainability Checklist

Project Management Toolkit — Sustainability Checklist

Project:	<insert project title>	Date:	<insert date>

Project vision

<Insert brief description of the vision of sustainability>

Sustainability review information

Previous sustainability review	<insert date and review number>	This sustainability review	<insert date and review number>
Project representative	<insert name>	Customer representative	<insert name>

Sustainability checks

Check number	Check	Target (sustained change)	Last review	This review
<insert check number>	<insert a check which will support the continued sustainability of the change which has been implemented> <note that sustainability checks do not track benefits>	<input a measurable target with appropriate units>	<insert measure, or not applicable if this is the first review>	<note level of measure and insert any appropriate comments>
<insert check x>	<insert as few or as many checks as necessary to assure all stakeholders that the change has been sustained>			

Summary comments and next steps

<insert comments regarding the results of the sustainability check and any actions agreed with the customer>

Is the change completely sustained?	Yes/No <delete as appropriate>	Date of next sustainability check	<insert date>

Appendix 9-9 – RACI Chart

Project Management Toolkit – RACI Chart

Project:	<insert project title>		Date:	<insert date>
Names → **Activity ↓**	<insert Project Team member name>	<insert Project Team member name>	<insert stakeholder name>	<insert stakeholder name>
<insert activity>	R	A	C	I
<insert activity>	Responsible	Accountable	Consulted	Informed
<insert activity>	<decide who will complete each activity>	<decide who is accountable for the activity completion>	<decide who needs to be consulted for the activity tobe completed>	<decide who needs to be informed of the results>

Appendix 9-10 – Table of Critical Success Factors (CSFs)

Project Management Toolkit – Table of CSFs

Project:	<insert project title>		Date:	<insert date>

Critical path of success

<insert a brief description of the intended project outcome which requires the completion of the Level 1 critical success factors (CSF1)>

CSF definition				
Scope area (CSF Level 1)	**Objective tracking metric (CSF Level 2)**	**Critical milestone (CSF Level 3)**	**Accountable for CSF Level 3 delivery**	**Priority (within scope area)**
<insert CSF Level 1>	<insert the next level of scope which is required to deliver CSF1>	<insert the next level of scope which is required to deliver CSF2>	<who in the team is responsible for CSF3>	<how important is this objective vs other objectives – need to prioritize>
<insert specific success criteria>	<note that a CSF1 may have a number of associated CSF2s>	<note that a CSF2 would usually have a limited number of associated CSF3s>	<insert name>	<insert priority>
<insert CSF Level 1>	<insert CSF Level 2>	<insert CSF Level 3>	<insert name>	<insert priority>
	<insert CSF Level 2>	<insert CSF Level 3>	<insert name>	<insert priority>
	<insert CSF Level 2>	<insert CSF Level 3>	<insert name>	<insert priority>

Appendix 9-11 – Project Scorecard

Project Management Toolkit – Project Scorecard

Project:	*<insert project title>*	Date:	*<insert date>*

Project status summary

Project status	*<insert summary comments on the project status, for example current stage in the project roadmap, stakeholder or team summary>*
Project control status	*<insert summary comments on the overall risk profile, overall cost and schedule forecasts and any plans which need to be put in place to regain or maintain control>*

Key activities this period	Achievements
<insert the most important activities completed in this reporting period – activities started or finished, stage gates passed>	*<insert the most significant milestone achievements or CSF status>*

Key activities next period	Critical issues
<insert the most important activities to be completed in the next reporting period – what are the critical milestones which need to be tracked to keep driving the project forwards>	*<insert risks which are becoming potential issues to the overall project success, for example related to funds, resources, skills, people, time>*

CSF tracking

CSF Level 1	Progress		Risk rating	
CSF a	*<insert Red, Amber, Green (RAG) rating>*	*<insert text>*	*<insert risk rating>*	*<insert text>*
CSF b	*<insert RAG rating>*	*<insert text>*	*<insert risk rating>*	*<insert text>*
CSF c	*<insert RAG rating>*	*<insert text>*	*<insert risk rating>*	*<insert text>*
Overall project	*<insert overall RAG rating>*	*<insert text>*	*<insert overall risk rating>*	*<insert text>*

Appendix 9-12 – Risk Table

Project Management Toolkit – Risk Table

Project:						
<insert project title>			**Date:**	*<insert date>*		
Risk description			**Risk assessment**		**Action planning**	
Risk number	**Risk description**	**Risk consequence**	**Occur?**	**Impact?**	**Mitigation plan**	**Contingency plan**
Number all risks for use in risk matrix	Describe what could happen	Describe the consequence of the risk	Assess if low, medium or high probability of risk occurring	Assess if low, medium or high impact on project goal	Describe the actions which would either stop the risk occurring or minimize the impact	Describe the contingency plan which would minimize impact if the risk occurred
<insert risk number>	*<describe the risk scenario>*	*<describe the impact – what happens?>*	*<insert risk probability>*	*<insert risk impact>*	*<describe the mitigation action plan>*	*<describe the contingency action plan>*
<insert risk x>						

Appendix 9-13 – Risk Matrix

Project Management Toolkit – Risk Matrix

Project:	<insert project title>	Sponsor:	<insert name of assigned sponsor>
Project Manager:	<insert name of assigned Project Manager>	Date:	<insert date>

Probability		
	Typical scoring	Scoring for this project
Low	<50% chance	
Medium	50% chance	
High	>50% chance	

Impact		
Low	Minimal impact	
Medium	Significant impact, difficult to achieve desired outcome	
High	Critical impact, unable to achieve desired outcome	

Red risks

Amber risks

Green risks

Note 1: Probability scoring – only use more accurate probabilities IF you have the data to back this up.

Note 2: Impact scoring – it is important that the impact is defined in terms of impact on a specific CSF or the vision of success.

Appendix 9-14 – Glossary

Benefits criteria
The reason the project is being done; the articulation of benefits as linked to an organizational objective.

Benefits hierarchy (from *Project Management Toolkit*)
A tool which confirms the alignment of the intended project scope to the targeted business benefits.

Benefit metrics
Those measures which will confirm, after project delivery, that the business is realizing the benefits.

Benefits mapping
A method which identifies and articulates the benefits that relate to the specific organizational goal.

Critical success factors (CSF)
A quantifiable/measurable and identifiable action/activity that has the potential to impact the overall success of the project.

Critical path of success (from *Project Management Toolkit*)
The linkage of high-level CSFs to form a critical path of quantifiable/measurable actions/activities that have the potential to impact the overall success of the project.

Critical success hierarchy (from *Project Management Toolkit*)
The hierarchy of CSFs which define the scope of the project; with each level in the hierarchy the scope detail increases.

HVAC – Heating, Ventilation and Air Conditioning.

SMART – Specific, Measurable, Achievable, Relevant, Timely.

P&IDs – Piping and Instrumentation Diagrams.

FMEA – Failure Mode Effect Analysis – a risk assessment technique.

Gowning policy – A procedure to require specific clothing in order to work within an environmentally controlled area.

Appendix 9-15 – References

Goldratt, E.M. *Critical Chain*, The North River Press, Great Barrington, MA, USA (1997).

Kaplan, R.S. and Norton, D.P. *The Balanced Scorecard*, HBS Press, Boston, MA, USA (1996).

Melton, T. *Project Management Toolkit: The Basics for Project Success*, Elsevier (2007), 2nd Edition, Oxford, Great Britain.

Adams, A., Kiemele, M., Pollack, L. and Quan, T. *Lean Six Sigma: A Tools Guide*, Air Academy Associates (2004), 2nd Edition, Colorado Springs.

Cockman, P., Evans, B. and Reynolds, P. *Consulting for Real People*, McGraw Hill, Maindenhead, Great Britain (1999).

Beblin, R.M., *Management teams: Why they succeed or fail*, Elsevier (2004), 2nd Edition, Oxford, Great Britain.

Beblin, R. *Management Teams*, Elsevier (2006), 2nd Edition, Oxford, Great Britain.

Appendix 9-16 – Project Management Skills Toolkit Assessment

Project Management Skills Assessment

Project Manager:		*\<insert name\>*	Date:	*\<insert date\>*	Assessment By:	*\<insert name\>*	
Toolkit		Tool	Tool description	Level of skill	Score	Development plan
Basic	1	'Why?' Checklist	A set of checks which ensure that the reason 'why' a project is required is complete	*\<insert level of skill score 1 to 5\>*	*\<insert summary score for all tools in this toolkit and this stage\>*	*\<insert actions to develop a higher score\>*
	2	'How?' Checklist	A set of checks which ensure that a robust project delivery strategy is in place			
		Project Gantt Chart	A bar chart to show the time when project tasks will occur			
		RACI Chart	A chart which allows identification of specific roles and responsibilities within a team			
		Simple risk assessment	A method to collect and assess project risk data (likelihood and impact)			
		Project organization chart	An organogram showing all key team members and the internal and external reporting links			
		Simple cost plan	A method to collect and present data on the total cost of a project			
	3	'In Control?' Checklist	A set of checks which ensure that a project in progress is in control			
		Action logs	A method to log, plan and track action completion			
		Activity lists	A method to list, plan and track activity completion			
		Project progress report	A succinct report to indicate status on cost, scope and time project objectives			

Appendix 9-16 (Continued)

Project Management Skills Assessment

Project Manager:			Date:		Assessment By:		
<insert name>			<insert date>		<insert name>		
Toolkit		Tool	Tool description		Level of skill	Score	Development plan
Basic	4	'Benefits Realized?' Checklist	A set of checks which ensure that all project benefits have been realized				
		Project after action review	A project review activity which reflects on the journey, the outputs and what has been learnt				
		Project completion checklist	A set of checks which ensure that all scope is complete and ready for handover				
Standard	1	Business Case Tool	A document which succinctly shows the costs (money, assets, people, etc.) versus the benefits of a proposed project				
		Benefits Hierarchy	A method to demonstrate alignment of a project or idea for a project with an organizational goal, considering the main project deliverables and key benefit metrics				
		Path of CSFs	The linkage of high-level CSFs to form a critical path of quantifiable/measurable actions/activities that have the potential to impact overall project success				
	2	Project Charter	The customer agreed document which details the project scope and deliverables required to achieve the business benefits				
		Control Specification Table	An activity to select the chosen methodology and/or tool for control of scope, time and cost, including change management and progress measurement				
		Table of CSFs	Development of the scope and project objectives via detailed development of the CSFs				
		Work breakdown structure	A hierarchy of project activities required to complete the project				

(continued)

Appendix 9-16 (Continued)

Project Management Skills Assessment

| Project Manager: | <insert name> | Date: | <insert date> | Assessment By: | <insert name> |

Toolkit		Tool	Tool description	Level of skill	Score	Development plan
Standard	2	Logic linked Gantt or PERT Charts	A chart showing all the project activities with clear dependencies between activities and an analysed critical path			
		Resource histogram	Based on the data in the resource plan, this is a visual representation of the level of loading of each resource in the Project Team			
		Risk Table and Matrix	Tracks risks in CSF categories, assessing likely impact and probability and then defines mitigation and/or contingency plans as necessary			
		Project Delivery Plan	A complete project strategy incorporating 11 distinct planning themes			
		Deliverables matrix	A table showing document ownership, review and approval process, completion plan and status			
	3	Milestone Chart	A table of milestone dates for activities on or near the critical path of a project			
		Risk assessment and risk logs	Tracks risks as they actually occur or are avoided, the actions put into place as a result and the impact analysis (cost, time, etc.)			
		Resource chart	A timeline chart which defines resource type and quantity vs activity and time			
		Tracking RACI Chart	A progress tracking tool to check that the appropriate people are completing project activities			

Appendix 9-16 (Continued)

Project Management Skills Assessment

Project Manager:		<insert name>	Date:	<insert date>	Assessment By:	<insert name>	

Toolkit		Tool	Tool description	Level of skill	Score	Development plan
Standard	3	Change control log	A method to log proposed project changes and to state and track impact assessment, approval and completion			
	4	Benefits Tracking tool	A tool to track the delivery of benefits to predetermined targets			
		Project Handover tool	Confirmation that the project scope, objectives and CSFs have been delivered and accepted			
Expert	1	Benefits mapping	A method of mapping an organizational vision through objectives, to clear benefit criteria or categories			
		Benefits scoring	Assigns scores to financial and non-financial benefit metrics to compare total project benefit			
	2	Contract selection matrices	A list of criteria to assess, select and manage an external supplier			
		Communications and Stakeholder Management Plans	The assessment of stakeholders and development of key messages to communicate to support project success			
		Resource loaded Gantt Charts	A detailed resource loading plan (resource quality and quantity)			
		Schedule risk assessment	Determines schedule contingency for each item based on risk. Use to calculate and position schedule buffer			
		Schedule buffer	Based on critical path analysis and schedule risks. Removes all schedule contingency from individual activities and places them in buffers at specific points in the schedule based on risks to project completion			

(continued)

Appendix 9-16 (Continued)

Project Management Skills Assessment

Project Manager:	<insert name>	Date:	<insert date>	Assessment By:	<insert name>		

Toolkit		Tool	Tool description	Level of skill	Score	Development plan
Expert	3	Earned Value tool	Assesses project progress against tangible deliverables enabling assessment of forecast to completion			
		Schedule contingency rundown	Plans when schedule contingency should be used and tracks progress of actual use. Based on schedule buffer calculations			
		Cost contingency rundown	Plans when cost contingency should be used and tracks progress of actual use. Based on cost risk calculations.			
	4	Benefits Scorecard	A summary report showing how project-related benefit metrics link to organizational goals in the four areas of the balanced scorecard			
		Sustainability Checklist	A table of metrics tracked after project handover to demonstrate that changes have been sustained			

Scoring – for level of use

1 – Aware but no skill	2 – Have used but need support	3 – Can use competently with minimal support	4 – Can use competently without support	5 – Expert user can coach and support others in use

Scoring – overall

<65% – Basic	65% to 80% – Standard	>80% – Expert

Summary

<insert comments after completion – comment on whether the user has a toolkit rating consistent across the four project stages or whether the skill in any one stage is higher or lower>

Index